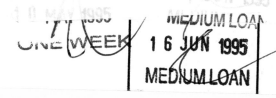

Marko Zlokarnik

Dimensional Analysis and Scale-up in Chemical Engineering

With 51 Figures

Springer-Verlag
Berlin Heidelberg New York
London Paris Tokyo
Hong Kong Barcelona Budapest

Professor Dr.-Ing. Marko Zlokarnik
Bayer AG
Geschäftsber. Zentrale Forschung
Gebäude B210
W-5090 Leverkusen, Bayerwerk
Germany

ISBN 3-540-54102-0 Springer-Verlag Berlin Heidelberg NewYork
ISBN 0-387-54102-0 Springer-Verlag NewYork Berlin Heidelberg

Library of Congress Cataloging-in-Publication Data
Zlokarnik, Marko,
Dimensional analysis and scale-up in chemical engineering / Marko Zlokarnik.
Includes bibliographical references.
ISBN 3-540-54102-0 (Springer-Verlag Berlin Heidelberg NewYork : acid-free paper).--
ISBN 0-387-54102-0 (Springer-Verlag NewYork Berlin Heidelberg : acid-free paper)
1. Chemical plants--Pilot plants. 2. Dimensional analysis.
I. Title.
TP155.5.Z56 1991
660--dc20 91-30861

Typesetting: Camera-ready by author
Offsetprinting: Color-Druck Dorfi GmbH, Berlin; Bookbinding: Lüderitz & Bauer, Berlin
51/3020-543210 Printed on acid-free paper.

This booklet is dedicated
to my friend and teacher
Dr. phil., Dr.-Ing. h.c.
Juri Pawlowski

Preface

Today chemical engineers are faced with many research and design problems which are so complicated that they cannot be solved with numerical mathematics. In this context, one only has to think of processes involving fluids with temperature-dependent physical properties or non-Newtonian flow behavior or fluid mechanics in heterogeneous material systems exhibiting coalescence phenomena or foaming. Scaling up of equipment for handling such material systems often presents serious problems which can frequently be overcome only by the aid of partial similarity.

In general, the university graduate has not been skilled enough to deal with such problems at all. On the one hand, treatises on dimensional analysis, the theory of similarity and scale-up methods included in common, run-of-the-mill textbooks on chemical engineering are out-of-date and only very infrequently written in a manner which would popularize these methods. On the other hand, there is no motivation for this type of research at universities since, as a rule, they are not confronted with scale-up tasks and therefore are not equipped with the necessary aparatus on the bench-scale.

All this mediates the totally wrong impression that the methods referred to are - at most - of marginal importance in practical chemical engineering, because they would have been otherwise dealt with in greater depth at university!

The aim of this booklet is to remedy this deficiency. It presents dimensional analysis in such a way that it can be immediatelly and easily understood, even without a mathematical background. Examples illustrating points which are currently of interest are used to explain each problem considered in the booklet.

However, the most extensive section of this booklet (2/3 of the total content) is devoted to the integral treatment of problems from the fields of mechanical and thermal unit operations and of chemical reaction engineering using dimensional analysis. In this respect, the term "integral" is used to indicate that, in the treatment of each problem, dimensional analysis was applied from the very beginning and that, as a consequence, the performance and evaluation of tests were always in acordance with its predictions.

A thorough consideration of this approach not only provides the reader with a practical guideline for his own use; it also shows him the unexpectedly large advantage offered by these methods.

I have been working with methods relating to the theory of similarity for more than thirty years and in all this time my friend and collegue *Dr. Juri Pawlowski* has been an invaluable teacher and adviser. I am indebted to him for numerous suggestions and tips and for his comments on this manuscript. I would like to express my gratitude to him at this point.

Finally, my sincere thanks also go to my employer, BAYER AG, Leverkusen, a company which has always permitted me to devote a considerable amount of my time to basic research in the field of chemical engineering in addition to my company duties and corporate research.

Contents

C Examples from the Field of Chemical Reaction Engineering

Introduction

The chemical engineer is generally concerned with the industrial implementation of processes in which chemical or microbiological conversion of material takes place in conjunction with the transfer of mass, heat, and momentum. These processes are scale-dependent, i.e. they behave differently on a small scale (in laboratories or pilot plants) and a large scale (in production). They include heterogeneous chemical reactions and most unit operations (e.g., mixing, screening, sifting, filtration, centrifugation, grinding, drying, and combustion processes). Understandably, chemical engineers have always wanted to find ways of simulating these processes in models to gain insights that will assist them in designing new industrial plants. Occasionally, they are faced with the same problem for another reason: an industrial facility already exists but will not function properly, if at all, and suitable measurements have to be carried out to discover the cause of the difficulties and provide a solution.

Irrespective of whether the model involved represents a "scale-up" or a "scale-down", certain important questions always apply:

1 How small can the model be? Is one model sufficient or should tests be carried out in models of different sizes?

2 When must or when can physical properties differ? When must the measurements on the model be carried out with the original system of materials?

3 Which rules govern the adaptation of the process parameters in the model measurements to those of the full-scale plant?

4 Is it possible to achieve complete similarity between the processes in the model and those in its full-scale counterpart?

These questions touch on the fundamental issues of the theory of models, a theory derived from the theory of similarity, which is, in turn, based on dimensional analysis. The following is a theoretical discussion of these questions, illustrated with a series of practical examples.

In today´s world of technological progress, it is becoming increasingly important to optimize processes both as a whole and in their individual steps and to do so with an eye to both economy and environmental protection. The days of generous estimates and consequent "safety margins" - described by *U.Grigull* as "ignorance margins" as long ago as in the early 1970´s - are definitely over. "All-round" plants, capable of manufacturing one product one day and another the next, and ultimately, after being tuned to the relevant process on site by means of the "black box method", capable of producing many times the intended output, are a thing of the past.

Attempting to optimize a process means looking more deeply into its physical and technical elements. The effects of each parameter that influences a technical process must be carefully examined. This involves answering increasingly complicated questions, for which mathematical solutions often do not exist. The researcher becomes more and more dependent on model experiments at a time when qualified laboratory personnel are scarce. Thus, problems of process technology have to be solved with a minimum of assistance as regards both finance and staff. In future research expenditures will probably be reduced still further while gleaning a maximum of information from the work performed.

In such a situation the theory of similarity may be of invaluable service. This theory facilitates the sensible planning and simple execution of experiments and the evaluation of the resulting data so as to produce reliable information on the size and process parameters of the large-scale plant, assuming, of course, that the method is applied correctly and as early as possible.

This treatise is divided into four sections.

Chapter One deals with the **fundamentals of dimensional analysis**. Their importance lies in the fact that, by way of the Π-theorem, they offer the only means of dealing with problems that cannot be formulated mathematically.

Chapter Two shows how a physical process can be described with a complete set of relevant physical quantities. It also demonstrates how this comprehensive pool of information serves as the source of a **set of dimension-**

less numbers which is usually much smaller than the set of the dimensional physical quantities and describes the problem just as completely.

Chapter Three discusses **similarity and scale-up**. It explains in detail why reliable scale-up is possible only within the framework of a complete set of dimensionless numbers. It also investigates the problems of partial similarity. It goes on to demonstrate how reliable scale-up can be obtained even where complete similarity is impossible because of the unavailability of small-scale substitutes for the physical properties of a full-scale system.

Chapter Four considers specific questions which arise if **substances with variable physical properties** (e.g., temperature-dependent or non-Newtonian) are to be treated by means of dimensional analysis.

When applying dimensional analysis to chemical engineering, the compilation of a complete list of problem-related physical quantities always causes the most difficulties. The **Appendix** includes a series of **practical examples** covering various questions from the fields of mechanical, thermal and chemical reaction engineering, which should provide assistance in handling this vital first step. Apart from being didactically useful, these examples are intended to serve as a "good performance certificate" for dimensional analysis and thus as an incentive to use it.

The following is a brief list of the advantages made possible by correct use of dimensional analysis. (These will become even clearer in the light of the subsequent examples!)

1. Reduction of the number of parameters required to define the problem. The Π theorem states that a physical problem can always be described in dimensionless terms. This has the advantage that the number of dimension*less* groups which fully describe it is much smaller than the number of dimensional physical quantities. It is generally equal to the number of physical quantities minus the number of basic units contained in them.

2. Reliable scale-up of the desired operating conditions from the model to the full-scale plant. According to the theory of models, two processes may be considered similar to one another if they take place under geometrically similar conditions and all dimensionless numbers which describe the process have the same numerical value.

3. A deeper insight into the physical nature of the process. By presenting experimental data in a dimensionless form, distinct physical states can be isolated from one another (e.g., turbulent or laminar flow region) and the effect of individual physical variables can be identified..

4. Flexibility in the choice of parameters and their reliable extrapolation within the range covered by the dimensionless numbers. These advantages become clear if one considers examples like the well-known Reynolds number, $Re = v\ l/\nu$, which can be varied by altering the characteristic velocity v, or a characteristic length l, or the kinematic viscosity ν. By choosing appropriate model fluids, the viscosity can be very easily altered by several orders of magnitude. Once the effect of the Reynolds number is known, extrapolation of both v and l is allowed within the examined range of Re.

1 Dimensional analysis

1.1 A brief historical survey

The end result of dimensional analysis is a complete set of dimensionless numbers that describe a physical process and that outline the conditions under which this process behaves "similarly" in the model and its full-sized counterpart; dimensional analysis is the basis of scale-up methods. *Lord Rayleigh* was aware of this when he referred to the studies in which he employed dimensional analysis as the study of similitude. Let us take this reference as our starting point in a historical survey of dimensional analysis which begins with the first attempts at scaling up a model, attempts made at a time when the very concept of dimensions was still unknown.

The desire to obtain information about a full-scale process by first carrying out tests on models has existed for centuries. *Leonardo da Vinci* [1] wrote,

> *"Vitruvius says that small models are of no avail for ascertaining the effects of large ones; and I here propose to prove that this conclusion is a false one."*

Galileo Galilei [2] later investigated the strength of mechanical parts of machines and *Isaac Newton* [3] clearly defined the concept of "mechanical similarity". *Bertrand* [4] formulated these rules unequivocally in 1847 and expressed Newton´s general law of similarity as a constraint which has to be satisfied by the four ratios of length, time, force and mass. However, the technical breakthrough in the field of similarity did not come until 1869/70, when *William Froude* [5] determined the drag on the ship by using model *experiments*, and 1883, when *Osborne Reynolds* [6] published the results of his model-based *experiments* on the flow of liquids through pipes. The basic works of *L.Prandtl*, *W.Nusselt*, *H. Gröber* and many others followed. The "science of models" was born. The significance of this method is well expressed by the famous words of *L.H.Baekeland* [7]:

"Commit your blunders on a small scale and make your profits on a large scale".[1)]

But how was it done? One demanded the "general physical similarity" of basic quantities (one spoke of geometrical, temporal, thermal and other forms of similarity) and used this similarity to formulate "scale-up rules" for secondary quantities such as velocities and rates of acceleration. The comparison of forces, energies, etc. led to the definition of the first dimensionless groups, which were later named and refered to as Newton´s, Reynolds´ or Froude´s "model laws". *Moritz Weber* comprehensively described these techniques of deriving dimensionless numbers in his two fundamental papers of 1919 and 1930 [A 1, A 2]. See also [A 3, Chap. 3].

These "numbers" can also be derived from the Navier-Stokes equations of hydrodynamics (cf. [A 3, Chap. 5] or [A 4]). The application of this method is limited because most of the processes of current interest do not lend themselves to mathematical description. Differential equations do, however, provide valuable information on the dimensionless numbers describing the process even if they cannot actually be solved (cf. the work of *Damköhler* [18], example C 1).

Based on the dimensional homogeneity of physical relationships, dimensional analysis provides a far more elegant approach to deriving dimensionless groups. The retrospect on the development of this basic method has to be corrected and supplemented since the publication of *H.Görtler*´s stu-dies in 1975 [8]. These amendments have been taken into account in this treatise.

In 1822 *Fourier* [9] coined the term "physical dimension" and emphasized that physical equations have to satisfy the property now described as dimensional homogeneity in relation to the chosen system of units. *H.von Helmholtz* [10] examined the dimensionless groups governing hydrodynamics and came up with the "Reynolds number" ten years before the corresponding publication by Reynolds appeared! *Lord Rayleigh* [11] was

---

1) The validity and the spirit of this remark are not diminished by the fact that Baekeland was not referring to true scale-up rules but rather to systematic experiments with single parts of small production units.

particularly influential in encouraging the use of and popularizing dimensional analysis.

In 1890 *A.Vaschy* [12] was the first to attempt to formulate the Π-theorem as a consequence of the dimensional homogeneity of physical equations. *A.Federmann* [13], however, proved the Π theorem in the course of a mathematical analysis of partial differential equations in 1911. Federmann´s work was not known to E.Buckingham who credited the discovery of the Π theorem to *D.Riabouchinsky* [14]. Posterity, on the other hand, gave all the credit to *E.Buckingham* because his paper of 1914 [15], in which he introduced the term Π theorem, finally aroused the scientists´ interest in this remarkable method. The theory states that a physical problem can always be described in dimensionless form with the advantage that the number of dimensionless groups required is equal to the number of dimensional parameters minus the number of basic dimensions involved. *P.W. Bridgman* [A 5] later demonstrated (1922) that this "rule of thumb" does not always apply. It is more correct to speak of the rank of a dimensional matrix rather than of the number of basic units.

The excellent textbooks of *P.W. Bridgman* (1922 and 1931) [A 5] and *H.L. Langhaar* (1951) [A 6] helped to popularize dimensional analysis and the theory of models derived from it. *R.C. Pankhurst´*s book [A 7] is another important English publication. Further attention must be drawn to the *L.I. Sedov´*s Russian textbook [A 8] which enjoys a good reputation. Of all contributions in German, *J.Pawlowski´*s book [A 9] is most worthy of mention. It deals with a series of questions related to the theory of similarity in an exact, mathematical fashion and is the only one to consider the treatment of variable physical properties (e.g., non-Newtonian viscosity) by dimensional analysis. The book also provides extremely simple guidelines for using matrix calculations to derive dimensionless numbers. (His technique will be referred to in detail later.) Pawlowski offers his own proof of the Π-theorem. *H. Görtler´*s book [A 10] includes a comprehensive, mathematical treatment of dimensional analysis and compares and comments on the lines of proof offered by several other authors.

1.2 Introduction to Dimensional Analysis

How should a physical problem be approached? We shall answer this question in considering three increasingly complex problems:

1. The period of oscillation of a pendulum;
2. The period of oscillation of small drops of liquid under the influence of their own surface tension;
3. The pressure drop of a fluid in a straight smooth pipe.

1. The period of oscillation of a pendulum

This problem is used in several treatises on dimensional analysis (e.g., in [A5]) as a simple but elegant means of demonstration.To establish what determines the period of oscillation of a pendulum it is common to begin with a list of all the variables that may be relevant to it, i.e., the *relevance list*. The dimensions of these variables (L length, M mass, T time) are then introduced. These variables should form a "physical relationship" that is independent of the system of dimensions chosen to measure them, a condition called *dimensional homogeneity.*

It may be assumed that the period of oscillation of a pendulum depends on the length and mass of the pendulum, the gravitational acceleration and the amplitude of swing:

Physical quantity	Symbol	Dimension
Period of oscillation	t	T
Length of pendulum	l	L
Mass of pendulum	m	M
Grav. acceleration	g	LT^{-2}
Amplitude (angle)	α	–

Our aim is to express t as a function of l , m , g and α :
$t = f(l, m, g, \alpha)$.

The functional relationship f must remain independent (invariant) of the choice of system of units. This self-evident requirement affects the structure of the relationship in a significant way. The numerical value given for time depends only on the size of the basic unit of time being used and is therefore always the same even if the basic units used for mass and length change. In order to ensure that the function remains unchanged even when the basic units for mass and length are changed, the corresponding values (containing M and L) in the function arguments have to be combined in such a way as to remain unaffected, i.e. these values have to be made dimensionless with regard to M and L.

The first thing that now becomes clear is that the basic unit for the mass M occurs only in the mass m itself. Changing this basic unit , e.g. from kilograms to pounds, would change the numerical value of the function. This is unacceptable. Either our list should have included a further variable containing M, or mass is not a relevant variable. If we assume the latter, the above relationship is reduced to:

$t = f\ (l, g, \alpha).$

Both l and g incorporate the basic unit of length. When combined as a ratio (l/g) they become dimensionless with regard to L and thus independent of changes in the basic unit of length:

$t = f\ (l/g, \alpha)$

Since the angle α has no dimension, we are left with the dimension T on the left-hand side of the equation and T^2 on the right. To remedy this, we will have to write $\sqrt{l/g}$:

$t = \sqrt{l/g}\ f(\alpha).$

This equation is the only statement that dimensional analysis can offer in this case. It is not capable of producing information on the form of f. The integration of Newton´s equation of motion for small amplitudes leads to $f = 2\pi$ and independent of α ; cf. [A 10]. The relationship can now be expressed as:

$$\boxed{t\sqrt{g/l} = 2\pi}$$

The expression on the left is a *dimensionless number* with a numerical value of 2π.

The elegant solution of this first example should not tempt the reader to believe that dimensional analysis can be used to solve every problem at his desk. To treat this example by dimensional analysis, the law of free fall had to be known. This knowledge was gained empirically by *G.Galilei* [2] in 1604 through his *experiments* with the inclined plane. *Bridgman's* [A 5] comment on this situation is particularly appropriate:

> *"The problem cannot be solved by the philosopher in his armchair, but the knowledge involved was gathered only by someone at some time soiling his hands with direct contact."*

2. The period of oscillation of small drops of liquid under the influence of their own surface tension

This is a slightly more complex problem than the period of oscillation of a pendulum. The drops should be unaffected by gravity and the oscillation should cause no more than periodical deformation (sphere to ellipsoid). The period of oscillation will then obviously depend only on the surface tension of the drop, its density and its diameter.

Physical quantity	Symbol	Dimension
Period of oscillation	t	T
Surface tension	σ	$M\,T^{-2}$
Density of the liquid	ρ	$M\,L^{-3}$
Diameter of the drop	d	L

Our aim is to find the relationship

$$t = f\,(\sigma, \rho, d),$$

which must be such that the numerical value for t is independent of the choice of the system of dimensions. By proceeding as in the first example, the equation can be made dimensionless with regard to M by forming $\sigma/\rho\ [L^3\,T^{-2}]$ and with regard to L via $\sigma/\rho\, d^3\ [T^{-2}]$. Dimensional equality with T can be achieved via $\sqrt{\rho\, d^3/\sigma}\ [T]$.

The result:

$$t = \text{const}\sqrt{\rho\, d^3/\sigma} \qquad \text{or} \qquad \boxed{t\sqrt{\sigma/\rho\, d^3} = \text{const}}$$

can be confirmed experimentally.

The question then arises of how we knew that t depended only on the physical properties ρ and σ and not on other factors such as the viscosity or the compressibility of the liquid. The only reason is that previous *experiments* showed that the role of viscosity is insignificant in nonviscous liquids, that compressibility is irrelevant when tiny drops are concerned and that the above conclusion is valid only under these conditions.

Bridgman has a provocative question to ask on this subject: "What use is all this dimensional analysis if we still need to resort to extensive preliminary experiments to solve the problem!?" The answer is simple. It is indeed true that, when compiling a reliable relevance list, comprehensive preliminary experiments are often required. However, the two above examples clearly show that dimensional analysis then makes it possible to discover physical laws that would otherwise have required further systematic experiments.

It must be noted here that the dimensional analysis presented in examples 1. and 2. can be performed in this simple manner only when the result is a single dimensionless group. The problems, however, with which one is confronted nowadays are much more complex and therefore require a more systematic approach. Example 3. introduces a method which is still presented in most textbooks as the standard method.

3. Pressure drop of a homogeneous fluid in a straight smooth pipe.

When studying the flow of a homogeneous fluid (e.g., water or air) in a straight smooth pipe, the pressure drop Δp must be a function of the pipe geometry (diameter d and length l), the physical properties of the fluid (density ρ and viscosity η) as well as its velocity v. The relevance list thus becomes:

$\{\Delta p; d, l; \rho, \eta; v\}$.

For the relationship between these quantities

$$f(\Delta p, d, l, \rho, \eta, v) = 0 \qquad \text{(a)}$$

to be dimensionally homogeneous, it must be possible to form a *dimensionless product* Π ($\Pi = 1$) with them:

$\Pi = \Delta p^{\alpha} d^{\beta} l^{\gamma} \rho^{\delta} \eta^{\varepsilon} v^{\zeta}$. (b)

$\Pi = [ML^{-1}T^{-2}]^{\alpha} [L]^{\beta} [L]^{\gamma} [ML^{-3}]^{\delta} [ML^{-1}T^{-1}]^{\varepsilon} [LT^{-1}]^{\zeta} = 1$ (c)

It follows for the three basic dimensions (mass M, length L, time T) :
for M : $\alpha + \delta + \varepsilon = 0$; (d)
for L : $-\alpha + \beta + \gamma - 3\delta - \varepsilon + \zeta = 0$; (e)
for T : $-2\alpha - \varepsilon - \zeta = 0$. (f)

Since only three equations with six unknowns are available, there is an infinite number of possible Π numbers. If, for example, α, β, and γ are considered as given, then the rest can be derived as follows.

By solving equation (d) for ε and substituting this in (e) and (f) we obtain:
$\beta + \gamma - 2\delta + \zeta = 0$ (e 1)
$-\alpha + \delta - \zeta = 0$ (f 1)

Elimination of ζ from these equations leads to:
$\delta = -\alpha + \beta + \gamma$
$\varepsilon = -\beta - \gamma$
$\zeta = -2\alpha + \beta + \gamma$

Equation (b) can now be rewritten as follows:

$\Pi = \Delta p^{\alpha} d^{\beta} l^{\gamma} \rho^{-\alpha + \beta + \gamma} \eta^{-\beta - \gamma} v^{-2\alpha + \beta + \gamma}$ or:

$$\boxed{\Pi = \left(\frac{\Delta p}{\rho v^2}\right)^{\alpha} \left(\frac{d \rho v}{\eta}\right)^{\beta} \left(\frac{l \rho v}{\eta}\right)^{\gamma}}.$$

The resulting dimensionless product Π thus consists of three dimensionless groups (numbers). The first is the so-called Euler number, $Eu \equiv \frac{\Delta p}{\rho v^2}$, the second the Reynolds number, $Re \equiv \frac{d \rho v}{\eta}$, and the third, when recombined with the second, the aspect ratio, L/d.

The pressure drop of a fluid in a straight smooth pipe can therefore be expressed in dimensionless form with the relationship (cf. p. 25 - 27 and Fig.1):

$\boxed{Eu = f(Re, L/d)}$

The above method originates from *Lord Rayleigh*, see e.g. [16]. *Langhaar* [A 6] points out that *Buckingham´*s method [15] differs only superficially from that of *Rayleigh*, leading to the same result. The various methods will not be discussed further, since they are presented in detail elswhere (see e.g. [A 6] and [A 10]). It should be noted, however, that the method which employs the dimensional matrix, discussed in detail in Chapt. 2 and used in all examples in the Appendix, is based on the same algebraic steps as employed in example 3.).

The three above examples clearly show how dimensional analysis deals with specific problems and what conclusions it allows. It should now be easier to understand *Lord Rayleigh´*s sarcastic comment with which he began his short essay on "The Principle of Similitude" [11]:

> *"I have often been impressed by the scanty attention paid even by original workers in physics to the great principle of similitude. It happens not infrequently that results in the form of "laws" are put forward as novelties on the basis of elaborate experiments, which might have been predicted a priori after a few minutes´ consideration".*

1.3 Fundamentals of Dimensional Analysis

1.3.1 *Physical quantities and the relationship between them*

When studying physical processes, one strives to obtain quantitative relationships between physical quantities which may or may not be of the same type. Types of physical quantities or *entities* (e.g. lengths, masses, times, etc.) are regarded as purely qualitative notions[2)]. They are in fact described by their method of measurement. A *physical quantity*, on the other hand, is a quantitatively described object (e.g., a mass of 5 kg). We must therefore discover which conditions allow physical appearances to

---

2) Notice that quantities of different types may, however, have the same dimensions. Examples: rotational speed – frequency – shear rate; kinematic viscosity – diffusivity – heat diffusivity.

be quantified and treated as mathematical objects for which functional relationships can then be formulated.

These questions touch on the principles of dimensional theory, which in turn deals with the rules of the formal language used to express physical statements. They are based on two important postulates:

1. *Any two physical quantities of the same type can be compared with one another by suitable means of measurement and a positive real number can than be assigned to this pair.*

If, for example, two bodies are weighed, the amount (numerical value) by which one body is lighter or heavier than the other can be determined. *Basic (or primary) quantities*, such as mass, length, or time, are thus quantified through comparison with their corresponding "standards". Once a standard – the basic unit – has been arbitrarily chosen, every quantity can be associated with a positive real number called its numerical value with respect to the chosen basic unit:

Physical quantity = numerical value $*$ basic unit

Example: a mass of 5 kg = 5$*$kg. (The symbol $*$ indicates that the combination is the result of a physical comparison. 5$*$kg is abbreviated as 5 kg for practical purposes.)

2. *Every physical interrelation can be represented as a relationship between the physical quantities involved.*

For example, the acceleration of a body can be described by a relationship R(F, m, a), which states that a specific force F is required to impart an acceleration a to a mass m. According to Newton's second law of motion, this relationship has the form $F = m\ a$. Numerical calculations with this equation, however, are still not possible as long as numerical values have not been assigned to these quantities. This can only be accomplished through a choice of arbitrary basic units and the quantification of the physical quantities involved, by comparison as outlined in postulate 1.

The second example includes two physical quantities, the force F and the acceleration a , which are termed *secondary or derived quantities* because they consist of a number of basic ones. The borderline separating primary and secondary quantities is largely arbitrary. Until recently, a system of dimensions (the so-called technical system) was used in which force was a primary dimension instead of mass.

Secondary quantities are not quantified by comparison with standards of the same type. Instead, relevant basic quantities are measured and the results combined according to specific rules. (Example: In order to obtain a numerical value for velocity, the distance and the time required to cover it are measured. The first of these numerical values is then divided by the second: In layman´s terms we would say that distance is divided by time.)

1.3.2 *Consistency of secondary units and invariance of physical relationships*

For physical relationships to be independent (invariant) of the choice of system of units, two requirements must be met: (a) All secondary units must be consistent[3)] with the basic units and (b) the relationships must be dimensionally homogeneous.

The first requirement means that in the SI system with the basic units {kg, m, s, ...} the velocity, for example, cannot be expressed in km/h but rather in m/s. The second requirement means that all additive terms in a relationship must have the same dimension and any function arguments (e.g., sin, exp) must be dimensionless.

1.3.3 *Physical dimensions, Systems of dimensions, Dimensional constants*

Physical laws definine relationships for secondary or derived quantities (e.g., velocity = length/time) and thus secondary units.
Example: Newton´s second law states that the force F is given by $F = m$ [kg] $\times\, a$ [m s^{-2}]. From this we learn that the force has the unit of [kg m s^{-2}] and thus the dimension [M L T^{-2}]. In this way dimensions can be assigned to all physical entities, which then form their own *System of*

---

3) The consistency (coherence) of secondary units has been neither defined nor required so far; it has nevertheless been fulfilled in all the examples.

dimensions. The currently used "International System of Dimensions" SI (Système International d'Unités; Table 2) prescribed by the Geneva Convention in 1954, is based on the following *seven* base quantities and their corresponding base units, see Table 1. (c.f. ISO/DIS of January 1972: "SI units and recommendations for the use of their multiples and of certain other units".)

Base quantity	**Base dimension**	**Base unit**
Length	L	m
Mass	M	kg
Time	T	s
Thermodyn. temperature	Θ	K
Amount of substance	N	mol
Electric current	I	A
Luminous intensity	J	cd

Table 1: Base quantities, their dimensions and units according to SI

Some important secondary units have been named after famous researchers:

Force expressed in Newtons: $N = kg\ m\ s^{-2}$

Energy expressed in Joules: $J = N\ m = kg\ m^2\ s^{-2}$

Power expressed in Watts: $W = J\ s^{-1} = N\ m\ s^{-1} = kg\ m^2\ s^{-3}$

In building a System of Dimensions it is sometimes necessary to introduce dimensional constants in order to preserve the consistency of the system.

Example 1: According to Newton´s second law, force is expressed as $F \sim m\ a$ with the dimension $[M\ L\ T^{-2}]$. According to Newton´s third law, however, force is also given by $K \sim (m_1\ m_2)/r^2$, which would lead to the dimension $[M^2\ L^{-2}]$. This inconsistency is resolved by agreeing on $F = m\ a$ and postulating a dimensional ("gravitational") constant G , defined by the third law:

$K = G\ (m_1\ m_2)/r^2$ with the dimension $G\ [M^{-1}\ L^3\ T^{-2}]$.

This example shows that *the dimensional homogeneity of physical relationships and the invariance of their functional representations can always be preserved through the introduction of dimensional constants.*

Physical quantity	SI [MLTΘ]	[MLTΘH]
mass	M	
length	L	
time	T	
temperature	Θ	
amount of heat		H
volume	L^3	
angular velocity frequency, shear rate	T^{-1}	
velocity	$L\,T^{-1}$	
acceleration	$L\,T^{-2}$	
kinematic viscosity diffusion coefficient thermal diffusivity	$L^2\,T^{-1}$	
density	$M\,L^{-3}$	
moment of inertia	$M\,L^2$	
surface tension	$M\,T^{-2}$	
dynamic viscosity	$M\,L^{-1}\,T^{-1}$	
momentum	$M\,L\,T^{-1}$	
force	$M\,L\,T^{-2}$	
pressure, stress	$M\,L^{-1}\,T^{-2}$	
angular momentum	$M\,L^2\,T^{-1}$	
energy, work, torque	$M\,L^2\,T^{-2}$	
power	$M\,L^2\,T^{-3}$	
mechan. equiv. of heat J	1	$M\,L^2\,T^{-2}\,H^{-1}$
specific heat capacity	$L^2\,T^{-2}\,\Theta^{-1}$	$M^{-1}\,\Theta^{-1}\,H$
thermal conductivity	$M\,L\,T^{-3}\,\Theta^{-1}$	$L^{-1}\,T^{-1}\,\Theta^{-1}\,H$
heat transfer coefficient	$M\,T^{-3}\,\Theta^{-1}$	$L^{-2}\,T^{-1}\,\Theta^{-1}\,H$
Stefan-Boltzmann-const.	$M\,T^{-3}\,\Theta^{-4}$	$L^{-2}\,T^{-1}\,\Theta^{-4}$

Table 2: Often used physical quantities and their dimensions according to the currently used SI (left) and the dimensional system in which the amount of heat was a basic quantity with the dimension of H (right).

Example 2: *Mechanical* energy is defined by the relationship: Energy = Force × Distance, leading to the dimension [M L^2 T^{-2}]. On the other hand, *thermal* energy is given by *Robert Mayer* as $E = J\,Q$. Here the amount of heat Q is a basic quantity with the primary dimension [H] and the primary unit kcal. The dimensional constant *J*, the so-called "*Joule´s* mechanical equivalent of heat" has the dimension [M L^2 T^{-2} H^{-1}] and the value of e.g. J = 4200 kg $m^2/(s^2$ kcal). An existing system of dimensions can be altered by making a dimensional constant dimensionless. Example: By making the mechanical equivalent of heat *J* equal to a dimensionless constant with the value of one, the heat Q is given the dimension of energy [M L^2 T^{-2}]. The system of dimensions {M L T Θ H} is thus reduced to the SI system {M L T Θ}, in which the calorific quantities such as c, λ, k and α now have new dimensions, see Table 2.

This illustrates that dimensions are not inherent properties of physical quantities, but merely expressions of the laws which define them.

1.3.4 *The dimensional matrix and its linear dependence*

Practical application of dimensional analysis leads to the solution of homogeneous, linear equations of the sort already discussed in example 3 on pages 11 - 13. Mathematical calculations with determinants (quadratic arrangements of numbers) or with matrices (right-angled arrangements of numbers) can be used for this purpose.

Our example is the problem involving velocity v , length l , mass m , density ρ , viscosity η and gravitational acceleration g. We begin by assembling a dimensional matrix in which the dimensions of these physical quantities are arranged as in a table (Remember: the dimension of g is [M^0 L^1 T^{-2}]):

	v	l	m	ρ	η	g
M	0	0	1	1	1	0
L	1	1	0	−3	−1	1
T	−1	0	0	0	−1	−2

This matrix consists of 3 "rows" and 6 "columns".

For the next step we need to know whether the rank, r, of this matrix really is three. It is important to know whether the rows of the matrix are linearly independent of each other or not. If one of the rows is simply any sum of the two others, then it can be described as a linear combination of these two and is dependent on them. In this case the rank of the matrix is lower than the number of rows it contains.

The rank of the matrix will not change if the rows that constitute a linear combination are eliminated. The rank of the matrix is normally given by the order of determinants; this can be found out particularly easily with the aid of the so-called *Gaußian* algorithm (as suggested by *Pawlowski* [A 9]). The dimensional matrix must exhibit a zero-free main diagonal and all the places below it must consist exclusively of zeros. The number of elements of the main diagonal which do not disappear but form a continuous sequence is the rank, r, of the matrix.

In practice the structural elements of the dimensional matrix (here the dimensions of the physical quantities) are arranged in such a way as to fulfill this requirement. If necessary, the columns can also change places (as in our example):

	m	l	v	ρ	η	g
M	1	0	0	1	1	0
L	0	1	1	−3	−1	1
T	0	0	−1	0	−1	−2

If this method fails, we can use additional *equivalence transformations*. in which single rows or linear combinations of rows are added together. We shall illustrate this simple method with a small but transparent example. It should be proven that the three physical quantities: density ρ, dynamic viscosity η and the kinematic viscosity ν are dependent linearly on one another, because they are bound together by the definition equation: $\nu \equiv \eta/\rho$. The rank of the dimensional matrix cannot be three but must be two. The proof is as follows:

	ρ	η	ν
M	1	1	0
L	−3	−1	2
T	0	−1	−1

	ρ	η	ν
M	1	1	0
1/2(3M+L)+T	0	1	1
T	0	−1	−1

	ρ	η	ν
M	1	1	0
1/2(3M+L)+T	0	1	1
1/2(3M+L)+2T	0	0	0

From the structure of the middle matrix we immediately recognize that elimination of -1 in the second column of the T row automatically eliminates -1 in the third column too. The last row then consists exclusively of zeroes (see the right matrix). This proves that the rank of this dimensional matrix is only two.

1.3.5 *The Π Theorem*

This theorem forms the basis for a discussion of physical relationships within the framework of the theory of similarity. It reads:

*Every physical relationship between **n** physical quantities can be reduced to a relationship between **m = n − r** mutually independent dimensionless groups, whereby **r** stands for the rank of the dimensional matrix, made up of the physical quantities in question and generally equal to the number of the basic quantities contained in them.*

The Π theorem can be proven in a number of different ways (e.g. see [A 10]). However, since we have chosen to use *Pawlowski´*s elegant and simple method [A 9] of deriving complete sets of dimensionless groups by the means of matrix calculation, his proof of the Π-theorem, which is based on the same procedure, will be presented here.

Let us examine a physical relationship involving x_k $(k = 1, 2, ... n)$ physical properties of any type. These quantities can be separated into two sets: the i-set and the j-set. The i - set contains only x_i quantities with dimensions which are linearly *in*dependent of one another; their units may be either primary or secondary ones. The j - set consists of all the other quantities x_j, the dimensions of which are linearly dependent on the i-set. Since every physical relationship consists at least of one quantity with a dimension that can be expressed in terms of the dimensions of the other quantities - a consequence of the dimensional homogeneity - each set must contain at least one element; it follows for $n \geq 2$ that neither r nor m is zero.

This classification of the physical quantities involved in the process in consideration is illustrated by the following dimensional matrix, <u>Fig.1</u>.

This matrix consists of a quadratic *unity matrix* and a *residual matrix*.

The unity matrix contains only x_i-quantities which are linearly independent of one another; its main diagonal is formed solely of ones and the remaining elements are all zero. The x_j-quantities form the residual matrix. Its elements p_{ij} depend on the dimensions of the x_j quantities with respect to x_i ones.

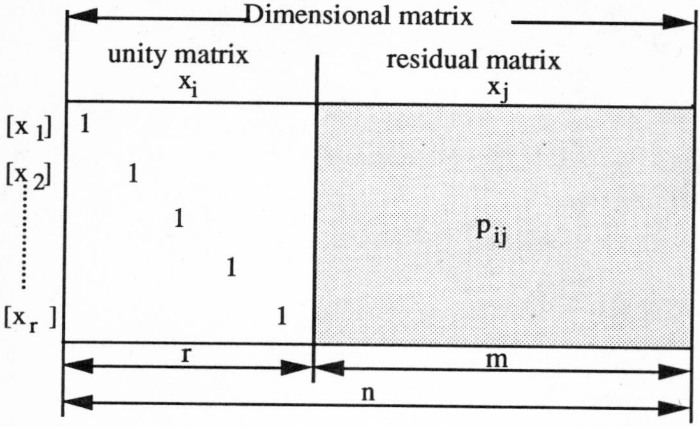

Fig.1: Dimensional matrix showing classification of physical quantities with dimensions that are linearly independent (x_i) or dependent (x_j) of one another

The physical relationship under consideration

$f_1(x_k) = 0$ with $k = 1, 2, ..., n$

can thus be expressed by the equation

$f_2(x_i, x_j) = 0$

Replacing the original primary units with new ones, which differ from them by a factor of $(1/a_i)$, results in both the change of the numerical values from x_i to $a_i x_i$[4)] and the substitution of the x_j-quantities by the expression:

$$\left(\prod_{i=1}^{r} a_i^{p_{ij}}\right) x_j.$$

4) By changing the system of units the numerical values and the units behave inversely proportional to each other.

It thus follows that

$$f_2 \{ a_i x_i, x_j \prod_{i=1}^{r} a_i^{p_{ij}} \} = 0.$$

This equation is valid for any positive values of a_i. If a_i values are chosen so as to obtain $a_i x_i = 1$, the equation will change to:

$$f_2 \{ 1, x_j \prod_{i=1}^{r} x_i^{-p_{ij}} \} = 0.$$

It is now obvious that

the relationship has shrunk from its original n arguments to only m arguments, whereby $\boldsymbol{m = n - r}$.

This being the case, we can progress straight to:

$$f_2 \{ x_j \prod_{i=1}^{r} x_i^{-p_{ij}} \} = \Phi(\Pi_j) = 0 .$$

These new variables

$$\boxed{\Pi_j = x_j \prod_{i=1}^{r} x_i^{-p_{ij}}}$$

are often called Π variables (because of the product symbol Π) or dimensionless numbers or groups.

In German technical literature the term criterion ("Kennzahl") is commonly used. It emphesizes the fact that the dimensionless numbers characterize specific conditions (e.g. different ranges in hydro- or thermodynamics). From here on the term *dimensionless number* or *group* will be used in preference to "Π variable".

The relationships between the dimensionless numbers will be termed *characteristics* , where they characterize specific properties of a peace of apparatus or a process (e.g. power characteristic $\mathrm{Ne} = f(\mathrm{Re})$ of a stirrer, or heat tranfer characteristic $\mathrm{Nu} = f(\mathrm{Re}, \mathrm{Pr})$ of a mixing vessel, or the pressure drop characteristic $\mathrm{Eu}\, l/d = f(\mathrm{Re})$ of a fluid in a pipe).

2 Description of a Physical Process with a Full Set of Dimensionless Numbers

2.1 The Relevance List for a Problem

All the essential ("relevant") physical quantities (variables, parameters) which describe a physical or technological interrelation must be known before this process can be described with a full set of dimensionless numbers. This demands a thorough and critical appraisal of the process being examined.

Langhaar [A 6] points out that this first step of naming the process parameters may often require a "philosophical insight" into natural phenomena. *Bridgman* [A 5] goes further still; when discussing the example of the period of oscillation of a pendulum, see page 8, he remarked that, when clearly defining the physical interdependence, it is sometimes necessary to carry out preliminary tests "by someone at some time soiling his hands with direct contact".

The application of dimensional analysis is indeed heavily dependent on the available knowledge. *Pawlowski* [5] outlines the following five steps (cf. also Fig. 2):

1. The physics of the basic phenomenon is unknown.
   → Dimensional analysis cannot be applied.
2. Enough is known about the physics of the basic phenomenon to compile a first, tentative relevance list .
   → The resultant Π set is unreliable.
3. All the relevant physical variables describing the problem are known.
   → The application of dimensional analysis is unproblematic.
4. The problem can be expressed in terms of a mathematical equation.
   → A closer insight into the Π relationship is feasible and may facilitate a reduction of the set of dimensionless numbers .
5. A mathematical solution of the problem exists.
   → The application of dimensional analysis is superfluous.

It must, of course, be said that approaching a problem from the point of view of dimensional analysis remains useful even if all the variables relevant to the problem are not yet known (case 2 above): The timely application of dimensional analysis may often lead to the discovery of forgotten variables or the exclusion of fakes (see example of pendulum, page 8).

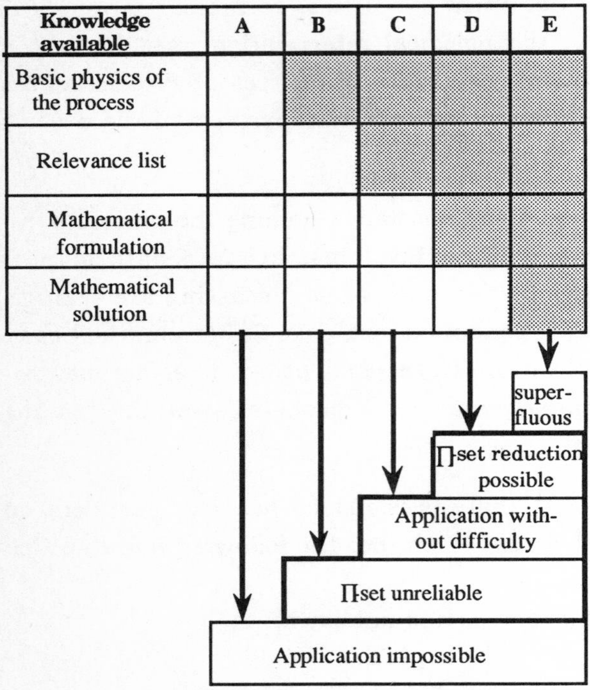

Fig. 2: Ease of application of dimensional analysis depending on the degree of knowledge available on a particular problem. (*Pawlowski* [5)])

The relevant physical variables comprise a single *target quantity* (the only dependent variable; this can also be a target function (e.g. residence time distribution) or even a target field (e.g., a temperature field) and a series of parameters which influence it. These parameters can be divided into three categories:

1 *geometric* variables
2 *material* parameters (*physical properties*)
3 *process-related* variables

---

5) Personal communication by J. Pawlowski, 1984

2.1.1 *Geometric variables*

Geometric variables are all measurements of length, radii of curvature, angles etc., which define the geometry of the problem under examination. When using dimensional analysis, however, it is sufficient to enter only one "characteristic" measurement of length l in the relevance list. This serves for the dimensionless expression of all the other geometrical parameters: $\Pi_{geom} = (l_i/l,\ r_j/l,\ \alpha_z$[6]$)$. The characteristic length chosen should have a particular significance for the process (see examples in Apendix).

When tests are carried out on a single piece of apparatus to which no geometric changes are made (e.g., on a model of an existing full-scale device), the relevance list need include only the characteristic length, but Π_{geom} = idem (= identical value) has to be recorded, since the Π relationship for this this particular piece of apparatus is, of course, functionally dependent on Π_{geom}. If, however, the intention is to vary some of the important geometrical variables during testing, these must be included in the relevance list. In this case, only the remainder of Π_{geom} has to be listed separately.

2.1.2 *Material parameters*

Material parameters include not only the *physical properties* of the system, e.g., viscosities, densities, concentrations, etc., but also pore volumes ε or the mass or volume fractions φ_i of the phases. Both of these last two variables are, by definition, dimensionless. It should also be noted that variables like kinematic viscosity, ν , must not be included in the problem relevance list, if the dynamic viscosity, η , *and* density, ρ , have already been listed, because ν is linked to these variables by the definition relation: $\nu \equiv \eta/\rho$ and therefore dependent on them (cf. the example on p. 19). It is, however, completely irrelevant which of the combinations, ρ and η , ρ and ν , or even η and ν is chosen. Other examples of definition relations between physical properties are: $\gamma \equiv \rho g$ for gravity and $a \equiv \lambda/\rho c$ for thermal diffusivity.

---

6) Angles are dimensionless geometrical parameters *per se*, because they are defined as ratios of two lengths.

2.1.3 *Process-related parameters*

The process-related parameters chosen should preferably be ones which can be measured directly and not derived ones; e.g., the throughput q of a liquid instead of its flow velocity, $v \sim q/A$, revolutions per minute n and not tip velocity, $u \sim n\, d$. It is obvious enough that it must be possible to measure any process parameter that is to be included in the relevance list: The relative velocity of solid particles in a dissolution process in a stirring vessel or that of the dispersed phases in a countercurrent extraction column are not measurable[7].

2.1.4 *Universal physical constants*

The relevance list must also include universal physical constants such as the universal gas constant, R, the speed of light in a vacuum, **c**, or even the acceleration of a gravitational field (on Earth the acceleration due to gravity, g), if these constants influence the process concerned.

> *The fact that a relevant physical quantity is a constant can never be a reason not to include it in the relevance list!*

By failing to consider the relevance of gravitational acceleration, g,the chemical engineer may find he has made a serious mistake!

Failing to consider gravitational acceleration when dealing with problems of process technology is clearly not new. Lord Rayleigh [11] complained bitterly saying:

> *"I refer to the manner in which gravity is treated. When the question under consideration depends essentially upon gravity, the symbol of gravity (g) makes no appearance, but when gravity does not enter the question at all, g obtrudes itself conspicuously".*

This is all the more surprising in view of the fact that the relevance of this quantity is easy enough to recognize if one asks the following question:

> *Would the process function differently if it took place on the moon instead of on Earth?*

If the answer to this question is yes, g is a relevant variable.

---

7) If these relative velocities were accessible to measurement, they could be interpreted as "intermediate quantities" and be preferably incorporated into the relevance list, because this would reduce the number of parameters, see section 2.1.5.

The gravitational acceleration g can be effective solely in connection with the density, i.e., in the form of gravity $g\rho$. When inertial forces play a role, the density ρ has to be listed additionally. Thus it follows that:

a) In cases involving the ballistic (Galilean) movement of bodies, the formation of vortices in stirring, the bow wave of a ship, the movement of a pendulum and other oscillation processes affected by the Earth's gravity, the relevance list comprises $g\rho$ *and* ρ.

b) Creeping flow in a gravitational field is governed by the gravity $g\rho$ alone.

c) In heterogeneous material systems with differences in density (sedimentation or buoyancy movements), the difference in gravity $g\Delta\rho$ *and* ρ play a decisive role.

2.1.5 *Intermediate quantities*

In some cases a closer look at a problem (or previous experience) facilitates reduction of the number of physical quantities in the relevance list. This is the case when some relevant variables affect the process by way of a so-called intermediate quantity. Assuming that this intermediate variable can be measured experimentally, it should be included in the problem relevance list, if this facilitates the removal of more than one variable from the list.

Example 1: Dispersion processes (e.g. emulsification or aeration) in heterogeneous, fluid material systems, e.g. liquid/liquid and gas/liquid in mixing vessels. In the turbulent flow range (η being irrelevant) the volume-related interphase surface, $a \equiv A/V$, depends on the diameter of the stirrer, d, the physical properties of the bulk phase (density, η, and surface tension, σ) and on the rotational speed n of the stirrer:

$$a = f(d, \rho, \sigma, n).$$

In the range of fully developed turbulence, a quasi-homogeneous material system exists. Here the variables n and d can be substituted by P/V (mixing power, P, volume of liquid, V) and thus reduce the number of variables by one:

$$a = f(P/V, \rho, \sigma)$$

Example 2: Homogenization of liquid mixtures with different densities and viscosities. The mixing time, θ, needed for a complete mixing of two Newtonian liquids, one resting in a layer on top of the other, in a vessel of known geometry (characteristic measurement of length is the stirrer diameter, d) depends on the rotational speed of the stirrer, n, on the physical properties of the two liquids ($\rho_1, \rho_2, \nu_1, \nu_2$), on their volume ratio, φ, as well as on the gravity difference $g\Delta\rho$:

$$\theta = f(d, \varphi; \rho_1, \rho_2, \nu_1, \nu_2 ; n, g\Delta\rho).$$

As mixing process progresses, it becomes clear that the premixing of the two liquids depends essentially on $g\Delta\rho$ while the final homogenization occurs within a material system, to which the physical properties of the uniform mixture apply: $\nu^* = f(\nu_1, \nu_2, \varphi)$ and $\rho^* = f(\rho_1, \rho_2, \varphi)$. The introduction of these intermediate physical quantities, ν^* and ρ^* , allows deletion of three variables from the relevance list! We can now write:

$$\theta = f(d; \rho^*, \nu^*; n, g\Delta\rho).$$

See [B 1] for further details.

Example 3: Dissolved air flotation makes use of the fact that tiny gas bubbles float the hydrophobic solid particles they have caught while desorbing. The relevant physical properties are only partially known and are not easy to measure. The problem can be circumvented by introducing the "floating" (surfacing) velocity of the particles as a "lumped parameter". This can be utilized as an intermediate quantity in the dimensional analysis - provided, of course, that it is determined separately in the same material system. See example A 6 on page 87 and [B 2] for further details.

2.2 Determination of a complete set of dimensionless numbers.

The relevance list represents the starting point for the determination of a complete set of dimensionless numbers. The process uses matrix calculation and consists of the following steps [A 9] :

a) construction of the *dimensional matrix*;

b) application of the *Gaußian algorithm*[8] in order to determine the rank r of the matrix (which may reduce the size of the matrix);

c) formation of the *unity matrix*;

d) formation of the *dimensionless numbers*;

e) possible *transformation* of these dimensionless numbers to provide more common (usually named) expressions, or such dimensionless groups which are more suitable for handling or describing the problem.

The procedure for constructing the dimensional matrix and determining the complete Π set is demonstrated by using the example of

The pressure drop of the volume flow in a straight, smooth pipe (ignoring inlet effects).

Here, the relevance list consists of

1. the *target quantity*: pressure drop, Δp,
2. the *geometrical variables*: the diameter, d, and length, l, of the pipe,
3. the *physical properties*: the density, ρ, and viscosity, ν, of the fluid, and
4. the *process-related parameter*: the volume throughput, q:

$$\{\Delta p; d, L; \rho, \eta; q\}$$

When combined with the SI dimensional system the dimensional matrix takes shape as follows:.

	Δp	q	d	l	ρ	ν
M	1	0	0	0	1	0
L	−1	3	1	1	−3	2
T	−2	−1	0	0	0	−1
	Core matrix			Residual matrix		

The nature of the steps of the subsequent process makes this dimensional matrix less than ideal because it is necessary to know that each of the individual elements of the residual matrix will appear in only one of the dimensionless numbers, while the elements of the core matrix may appear as "fillers" in the denominators of all of them. The residual matrix should

8) The Gaußian algorithm makes it obvious, whether the physical quantities in the core matrix are linearly independent of each other or not.

therefore be loaded with essential variables like the target quantity and the most important physical properties and process-related parameters. Variables with an as yet uncertain influence on the process must also be included in this group. Should such variables later prove irrelevant, only the dimensionless number concerned will have to be deleted while leaving the others unaltered.

Since the *core matrix* has to be transformed into a *unity matrix*, the "fillers" should be arranged in such a way as to facilitate a minimum of linear transformations. The following reorganization of the above dimensional matrix achieves both of these aims:

	ρ	d	v	Δp	q	1
M	1	0	0	1	0	0
L	−3	1	2	−1	3	1
T	0	0	−1	−2	−1	0

Core matrix Residual matrix

The next step is the application of the *Gaußian algorithm* (zero-free main diagonal, only zeroes below):

	ρ	d	v	Δp	q	1	
Z_1	1	0	0	1	0	0	$Z_1 = M$
Z_2	0	1	2	2	3	1	$Z_2 = 3M + L$
Z_3	0	0	1	2	1	0	$Z_3 = -T$

The rank of the matrix is three, r = 3. Only one further linear transformation of the rows is now required to transform the core matrix into a *unity matrix*.

	ρ	d	v	Δp	q	1	
Z'_1	1	0	0	1	0	0	$Z'_1 = Z_1$
Z'_2	0	1	0	−2	1	1	$Z'_2 = Z_2 - 2\,Z_3$
Z'_3	0	0	1	2	1	0	$Z'_3 = Z_3$

Unity matrix Residual matrix

When *generating dimensionless numbers*, each element of the residual matrix forms the numerator of a fraction while its denominator consists of the fillers from the uniform matrix with the exponents indicated in the residual matrix:

$$\Pi_1 \equiv \frac{\Delta p}{\rho^1 d^{-2} \nu^2} = \frac{\Delta p\, d^2}{\rho \nu^2}\,; \quad \Pi_2 \equiv \frac{q}{\rho^0 d^1 \nu^1} = \frac{q}{d \nu}\,; \quad \Pi_3 \equiv \frac{l}{\rho^0 d^1 \nu^0} = \frac{l}{d}\,.$$

The dimensionless number Π_1 does not usually occur as a target number for Δp. It has the disadvantage that it contains the essential physical property, ν, which is already contained in the process number (which is where it belongs). This disadvantage can easily be overcome by appropriately combining the dimensionless numbers Π_1 and Π_2. This results in the well-known Euler number, $Eu \equiv \Pi_1 \Pi_2^{-2} = \frac{\Delta p\, d^4}{\rho\, q^2}$, which is often combined with $\Pi_3 \equiv L/d$ to obtain an intensively formulated target number[9]:

$$Eu\, d/L \equiv \Pi_1\, \Pi_2^{-2}\, \Pi_3 = \frac{\Delta p\, d^4}{\rho\, q^2}\, d/l.$$

These <u>transformations</u> show how the obtained dimensionless numbers can subsequently be reshaped by appropriate combinations into forms which correspond to common or named dimensionless groups or are particularly suitable for evaluating or describing test results.

The structure of the dimensionless numbers depends on the variables contained in the core matrix. The Euler number, obtained by combining Π_1 and Π_2 in the example above, would have been obtained automatically if ν and q had been exchanged in the core matrix:

	ρ	d	q	Δp	ν	l	
Z_1	1	0	0	1	0	0	$Z_1 = M$
Z_2	0	1	0	−4	−1	1	$Z_2 = 3M + L + 3T$
Z_3	0	0	1	2	1	0	$Z_3 = -T$

$$\Pi_1' \equiv \frac{\Delta p\, d^4}{\rho\, q^2} = Eu \qquad \Pi_2' \equiv \frac{\nu\, d}{q} \qquad \Pi_3' \equiv \frac{l}{d}$$

9) If one had considered earlier the fact that – by neglecting the inlet effects – Δp is proportional to l, $\Delta p/l$ could have been entered into the relevance list straight away.

All Π sets obtained from one and the same relevance list are equivalent to each other and can be mutually transformed at pleasure!

The dimensionless number, Π_2, is in fact the well-known Reynolds number, Re, even though it appears in another form here. We shall explain the structure that a dimensionless number must have in order to be called a Reynolds number. (The example is equally valid for all other named dimensionless groups.) The Reynolds number is defined as any dimensionless number combining a characteristic velocity, v, and a characteristic measurement of length, l, with the kinematic viscosity of the fluid, $\nu \equiv \eta/\rho$. The following dimensionless numbers are equally capable of meeting these requirements:

$$\mathrm{Re} \equiv v\frac{d}{\nu} \; \left(= \frac{4\,q}{\pi\,d^2}\frac{d}{\nu}\right) \sim \frac{q}{d\,\nu} \qquad \text{and} \qquad \mathrm{Re} \equiv n\,d\frac{d}{\nu} = \frac{n\,d^2}{\nu}$$

This method of compiling a complete set of dimensionless numbers makes it clear that the numbers formed in this way can contain neither numerical values nor any other constant. These appear in dimensionless groups only when they are established and interpreted as ratios on the basis of known physical interrelations.

Example: $\mathrm{Re} \equiv \pi\,n\,d^2/\nu$, where $\pi\,n\,d$ is the tip speed, or $\mathrm{Eu} \equiv \Delta p/(v^2\,\rho/2)$, where $v^2\,\rho/2$ is the kinetic energy.

Since such expressions are of the same value as the analytically derived ones, it is always necessary to present the definition!

In the above-mentioned case of the pressure drop of the volume flow in a straight pipe, this method of compiling a complete set of dimensionless numbers produces the relationship:

$$\boxed{f(\mathrm{Eu}\; d/l\,,\; \mathrm{Re}\,) = 0}$$

The information contained in this relationship is the maximum that dimensional analysis can offer on the basis of a relevance list, which we assumed to be complete. Dimensional analysis cannot provide any information about the form of the function f, i.e., the sort of Π relationship involved. This information can only be obtained experimentally.

2.3 The Π-relationship

In their famous study *Stanton and Pannell* [B3] evaluated the function f of the interrelation $f\,(\mathrm{Eu}\; d/l, \mathrm{Re}) = 0$ by measurements. Figure 3 shows the result of their work which demonstrates impressively the significance of the Reynolds number for pipe flow[10]. λ stands for the "friction" or "resistance" coefficient[11], which is defined as follows:

$$\lambda \equiv \frac{\Delta p}{(\rho/2)\, v^2} \frac{d}{l} = 2(\pi/4)^2 \frac{\Delta p\, d^4}{\rho\, q^2} \frac{d}{l} = 1.24\ \mathrm{Eu}\ d/l.$$

The drawn-in curve is valid for the laminar flow range ($\mathrm{Re} < 2300$) where the Π relationships is:

$$\lambda = 64\ \mathrm{Re}^{-1} \quad \text{or} \quad \lambda\, \mathrm{Re} = 64.$$

(This connection could have been clearly demonstrated had the authors chosen to present their test results in a double-logarithmical plot; this would have produced a straight line with a gradient of –1 in this range of Re!)

$\lambda \mathrm{Re}$ can be viewed as a new dimensionless number that does not include the physical property density, ρ .

> *It is only with the Π relationship that the relevance to the problem and the operational range of individual variables becomes clear.*

This example also shows that the Π set compiled on the basis of the relevance list does no more than define the maximum Π space, which may well shrink in a Π relationship like the one cited above.

---

10) Referring to this figure, B.Eck remarks in his book on Technical Fluid Dynamics (1961, p.123, in German): "If one represented – as it was once usual – l as a function of velocity v, one would obtain not a curve but a galaxy. Here, Reynolds´ law must strike even a beginner with an absolute clarity."

11) These denotations for Π numbers are misleading, because they veil the true character of the dimensionless numbers.

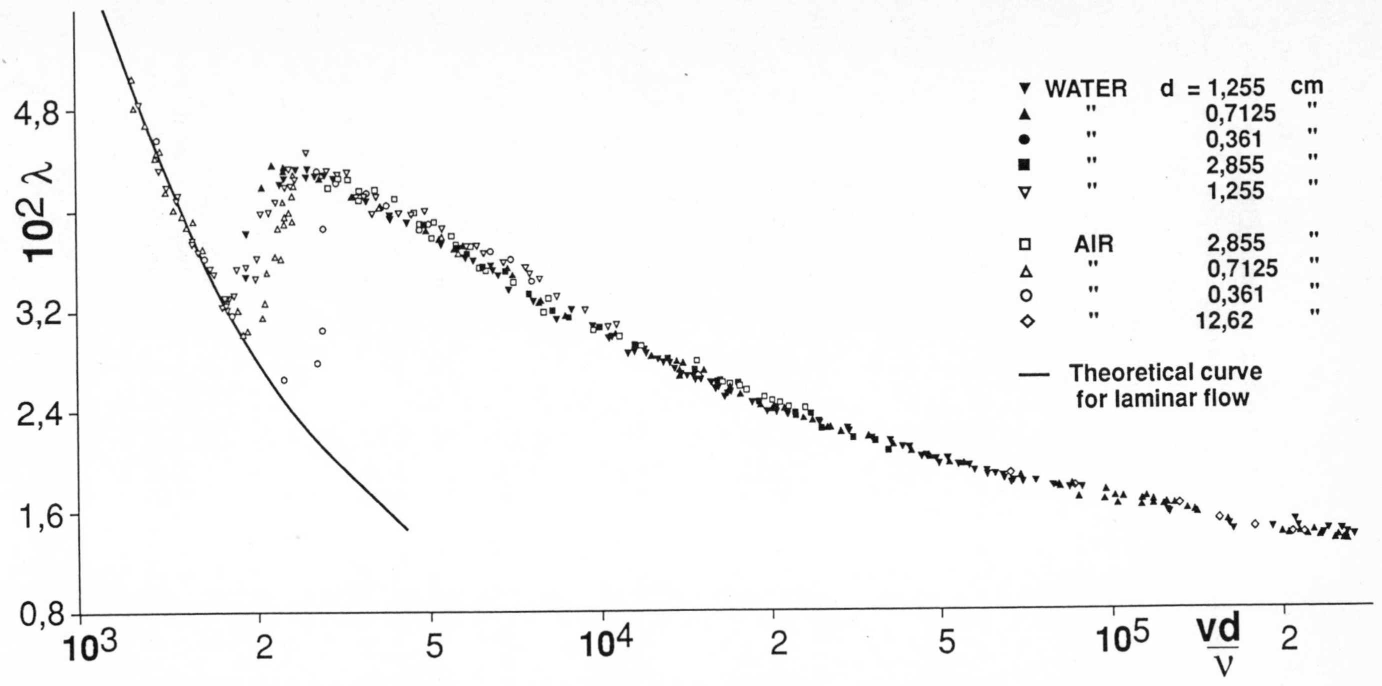

Fig. 3: The relationship λ(Re), the so-called *Pressure Drop Characteristic* of a smooth pipe in a single-logarithmic presentation. Measurements of *Stanton* and *Pannell* [B 3]

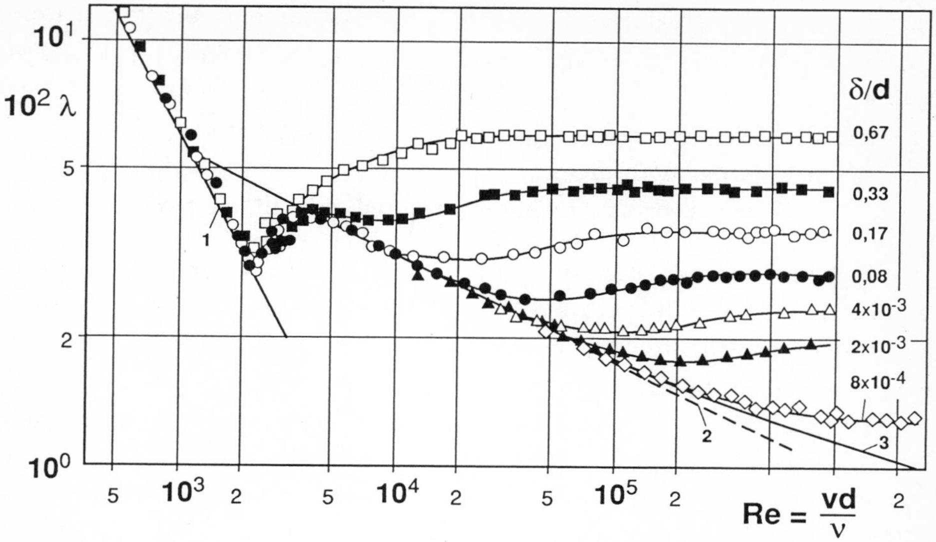

Fig. 4: Pressure drop characteristics λ (Re, δ/d) for a smooth and for a industrially rough pipe and for artificially roughened pipes. (See footnote 12!)

Straight line 1:	laminar flow:	$\lambda = 64\ Re^{-1}$
Straight line 2:	turbulent flow:	$\lambda = 0.3164\ Re^{-0.25}$ (*Blasius*)
Kurve 3:	Measurements of *Stanton* and *Pannell* (Fig. 3)	

$\delta/d = 2 \times 10^{-3}$ to 0.67 : Measurements of *Nikuradze* [B4]
$\delta/d = 8 \times 10^{-4}$: Measurements of *Galavics*

The following Π relationships are valid for the turbulent flow region ($Re > 2300$):

$\lambda = 0.3164\ Re^{-0.25}$ $\quad Re \le 8 \times 10^4$ (*Blasius*)

$\lambda = 0.0054 + 0.396\ Re^{-0.3}$ $\quad Re \le 1.5 \times 10^6$ (*Hermann*)

Figure 4 shows the same interrelation in a double-logarithmic presentation[12]. In this case an additional dimensionless geometrical parameter, wall roughness (δ/d = grain diameter/pipe diameter), is added to the

12) It should be noted here, that the two axes do not have the same scale: The ordinate has been stretched against the abscissa by a factor of 2.

"pressure drop characteristic" $\lambda(Re)$. (The measurements are taken from *Nikuradse and Galavics* [B4]). This variable was not included in the relevance list compiled on page 28 for two reasons. Its exclusion firstly allowed us to begin with the results of *Stanton and Pannell*, who carried out their tests in smooth pipes, and secondly it showed that any mistakes or omissions can be corrected when confronted with the test results!

The fact that the analytical presentations of the Π relationships encountered in engineering literature often take the shape of power products does not stem from certain laws inherent to dimensional analysis. It can be explained simply by the engineer´s preference for depicting his test results in double-logarithmic plots. Those sections of his curves that can be approximated as straight lines are then analytically expressed as power products. Where this proves less than easy, the engineer will often be satisfied with the curves alone, cf. Figs 3 and 4.

The "benefits" of dimensional analysis are often discussed. The above example provides a welcome opportunity to make the following comments. The five-parametrical dimensional relationship

$\{ \Delta p/l; d; \rho, \nu, q \}$

can be represented by means of dimensional analysis as $\lambda(Re)$ and plotted as a *single* curve (Fig. 3). If we wanted to represent this relationship in a dimensional way and avoid creating a "galaxy" at the same time (cf. footnote 10), we would need 25 diagrams with 5 curves in each! If we assumed that only 5 measurements per curve are sufficient, the graphic representation of this problem would still require 625 measurements. The enormous savings in time and energy made possible by the application of dimensional analysis are thus easy to appreciate. (These significant advantages were already pointed out by *Langhaar* [A6]).

2.4 Reduction of the size of the matrix

In Sect. 2.1 mention was made of *Buckingham´s* assumption, when defining the Π theorem in his famous study [15], that the number of Π numbers needed to create a complete picture of a problem can be derived from the number of dimensional variables minus the number of basic units of measurement included in them. *Bridgman* [A 5] corrected him, pointing out that this "rule of thumb" does not always work and that it would be more correct to refer to the rank of the matrix instead of the number of basic units of measurement.

Let us continue by demonstrating the above statement with a simple example. If two mutually miscible liquids with the same density and viscosity are mixed in a stirring vessel until the mixture is completely homogeneous, the mixing time, θ, will depend on a characteristic measurement of length (stirrer diameter, d) the physical properties of the mixture (density, ρ, diffusivity, $\mathfrak{D}$, and viscosity, ν) and the process parameter, the revolutional speed of the stirrer, n. The relevance list is therefore:
$\{\theta; d; \rho, \mathfrak{D}, \nu; n\}$.

The corresponding dimensional matrix is as follows:

	ρ	d	n	θ	ν	$\mathfrak{D}$
M	1	0	0	0	0	0
L	−3	1	0	0	2	2
T	0	0	−1	1	−1	−1

It is immediately clear that the basic unit of measurement for mass is present only in the density ρ. We must therefore delete column ρ and row M. The resultant matrix has the rank 2:

	d	n	θ	ν	$\mathfrak{D}$
L	1	0	0	2	2
−T	0	1	−1	1	1

The derivative Π set is:

$$\{ n\theta,\ Re \equiv n\, d^2/\nu,\ Sc \equiv \nu/\mathfrak{D} \}$$

Of course, this does not mean that density is an irrelevant quantity in this problem, it is simply already fully represented by the kinematic viscosity, $\nu \equiv \eta/\rho$. (Exchanging the kinematic viscosity, ν, for the dynamic viscosity, $\eta[M\, L^{-1}\, T^{-1}]$, would have resulted in the same three-parametrical Π set. In that case, however, the matrix would have retained its rank of $r = 3$.)

2.5 Change of dimensional systems

A change of dimensional system leading to an increase in the number of basic dimensions (e.g., adding the basic quantity of heat H[kcal] in heat transfer problems, cf. both dimensional systems in Table 2) may at first glance seem tempting for reducing the number of dimensionless groups for a given number of variables in the relevance list. It should be remembered, however, that expanding (or reducing) a dimensional system entails that the corresponding relevance list must also be expanded (or reduced) by the appropriate dimensional constant. This procedure, therefore, has no effect on the resulting number of Π variables. If, however, the dimensional constant is foressen to be irrelevant to the problem, it need not be added to the relevance list and the number of dimensionless numbers can be reduced by one. This often occurs in thermodynamic problems (cooling, heating, steady-state heat transfer) when mechanical heat generation is neglegible and *Joule*´s mechanical equivalent of heat J is therefore irrelevant. See Examples B1 and B3.

3 Similarity and scale-up

3.1 Basic Principles of Scale-up

In the Introduction it was pointed out that using dimensional analysis to handle a physical problem, and thus to present it in the framework of a complete set of dimensionless numbers, is a sure way of producing a simple and reliable scale-up from the small-scale model to the full-scale technical plant. The theory of models states that:

Two processes may be considered completely similar if they take place in similar geometrical space and if all the dimensionless numbers necessary to describe them have the same numerical value (Π_i = idem).

This statement is supported by the results shown in Fig. 3. The researchers carried out their measurements in smooth pipes with diameters d in the range of 0.36 - 12.63 cm. The physical properties of the fluid tested (water or air) varied widely. Nevertheless, every numerical value of Re still corresponded to a specific numerical value of λ! The Π relationship presented is thus valid not only for the examined laboratory devices but also for any other geometrically similar arrangement:

Every point in a Π framework, determined by the Π relationship, corresponds to an infinite number of possible implementations.

This characteristic of Π representation represents the basis of the concept[13)] of similarity based on dimensional analysis:

Processes which are described by the same Π relationship are considered similar to each other if they correspond to the same point in the Π framework.

Two realizations of the same physical interrelation are considered similar (*complete* similarity), when $m - 1$ dimensionless numbers of the m–dimensional Π space have the same numerical value (Π_i = idem), because

---

13) "Similarity mechanics"(Ähnlichkeitsmechanik), practised in Germany in the first half of this century, demanded that for complete similarity of two processes (apart from the geometrical similarity of both devices) all the ratios of forces, flows, etc, had to be equal; see [A1 & 2].

the m-th Π number will then automatically also have the same numerical value.

Let us attempt a scale-up with the simple example of Figure 3 by imagining a pipeline several hundred kilometers long, through which a given fluid (natural gas or crude oil) is to be transported with a given throughput . Our aim is to find the pressure drop of the fluid flow in the pipeline in order to design pumps and compressors.

We start by building a geometrically similar small-scale model of the technical pipeline. We already know the physical properties of the fluid, its throughput, and the dimensions of the technical plant and therefore have a numerical value for Re in operation. The same numerical value can be kept constant in the test apparatus by the correct choice of conveying device (pump, compressor, etc.) and model fluid. The pressure drop measured under these conditions will allow us to calculate the Euler number in the model. In this case the condition Re = idem automatically implies Eu = idem. The numerical value of the measured Euler number therefore corresponds to that of the full-scale plant. This then allows us to determine the numerical value of Δp in the technical plant from the numerical value of Eu in the model and the given operational parameters.

Of course, the concept of complete similarity does not guarantee that a process will be the same in the model and the full-scale version in *every* respect; it is only the same as regards the particular aspect under examination, which has been described by the appropriate Π relationship. In order to demonstrate this fact with the help of the above example, it should be remembered that the flow conditions in two smooth pipes of different scales should be considered similar when Re = idem and, according to the pressure drop characteristics, will therefore have the same numerical value of Eu d/l. This, however, does not mean that heat transfer conditions prevalent in the two pipes are the same; for that to be the case, the relevant Π relationship, Nu = f (Re, Pr), requires that both the Reynolds number and the Prandtl number have the same numerical value (temperature-independent physical properties of the medium being supposed).

The more comprehensive the similarity demanded between model and full-scale device and the greater the

scale-up factor $\mu \equiv l(\text{model})/l(\text{prototype, technical plant})$

the harder it is to perform the scale-up. It could even fail completely, if a material system with physical properties required for model experiments cannot be obtained (cf. Sect. 3.3). A further difficulty is that scale-up involving large changes of scale may cause changes to the Π space. An example here is the case of forced non-isothermal flow, in which progression in scale results in free convection and thus in the Grashof number, $Gr \equiv \beta \Delta T\, l^3\, g/\nu^2$ becoming relevant to the problem.

3.2 Experimental methods for scale-up

In the Introduction we were confronted with a number of questions often asked in connection with model experiments [19].

How small can a model be?
The size of a model depends on the scale factor, μ, and on the experimental accuracy of measurement. Where μ = 1:10, a 10 % margin of error may already be excessive. A larger scale for the model will therefore have to be chosen to reduce μ (cf. example B4, surface aeration).

Is one model scale sufficient or should tests be carried out in models of different sizes?
One model scale is sufficient if the relevant numerical values of the dimensionless numbers necessary to describe the problem (the "process point" in the Π space describing the operational condition of the technical plant) can be adjusted by choosing the appropriate process parameters or physical properties of the model material system. If this is not possible, the process characteristics must be determined in a series of models of different sizes, or the process point must be extrapolated from experiments in technical plants of different sizes (cf. Sect. 3.3).

When must model experiments be carried out exclusively with the original material ssystem?
Where the material model system is unavailable (e.g., in the case of non-Newtonian fluids) or where the relevant physical properties are unknown (e.g., foams, sludges, slimes) the model experiments must be carried out

with the material original system. In this case measurements must be performed in models of various sizes (cf. example A7).

The problems entailed in the unavailability of a model material system can occasionally limit the applicability of the theory of similarity. It would, however, be wrong to refer to "limits of the theory of similarity"!

3.3 Scale-up under conditions of partial similarity

When appropriate substances are not available for model experiments, accurate simulation of the working conditions of an industrial plant on a laboratory or bench scale may not be possible. Under such conditions, experiments with equipment of different sizes are customarily performed before extrapolation of the results to the conditions of the full-scale operation. Sometimes this expensive and basically unreliable procedure can be replaced by a well-planned experimental strategy, in which the process is devided into parts which are then investigated separately (e.g., prediction of the drag resistance of a ship´s hull using Froude´s approach, see Example 1 in this section) or by deliberately abandoning certain similarity criteria and checking the effect on the entire process (e.g., combined mass and heat transfer in a catalytic tubular reactor using *Damköhler´s* approach, see [18] and example C1).

Several "rules of thumb" for dimensioning different types of process equipment are, in fact, scale-up rules based unknowingly on partial similarity. These rules include the so-called *volume-related mixing power*, P/V, widely used for dimensioning mixing vessels, and the *superficial velocity*, v = q/A, which is normally used for scaling-up bubble columns. Some remarks on both these rules are given in Examples 2 and 3 at the end of this section, see also [20].

Example 1: *Drag resistance of a ship´s hull*
This problem represents the birth and the break-through of scale-up rules and is closely linked to the name of the brilliant researcher *William Froude* (1810 - 1879). Froude solved this significant scale-up problem with a clear physical concept and carefully executed experiments.

We shall first treat this problem by using dimensional analysis. The drag resistance F of a ship's hull of a given geometry (characteristic length, l, and displacement volume, V) depends on the speed, v, of the ship, the density, ρ, and kinematic viscosity, ν, of the water and, because of the bow wave formation, also on the acceleration due to gravity, g. The list of relevant quantities is thus:
$\{F; l, V; \rho, \nu; v, g\}$

The dimensional matrix

	ρ	l	v	F	ν	g	V
M	1	0	0	1	0	0	0
L	– 3	1	1	1	2	1	3
T	0	0	– 1	– 2	– 1	– 2	0

leads, after only two linear transformations, to the following unity matrix (with rank r = 3) and the residual matrix:

	ρ	l	v	F	ν	g	V	
M_1	1	0	0	1	0	0	0	$M_1 = M$
L_1	0	1	0	2	1	– 1	3	$L_1 = 3\,M + L + T$
T_1	0	0	1	2	1	2	0	$T_1 = -T$

The following four dimensionless numbers follow:

$\Pi_1 = F/(\rho\, l^2\, v^2)$ $\equiv$ Ne (Newton number)
$\Pi_2 = \nu / l\, v$ $\equiv$ Re^{-1} (Reynolds number)
$\Pi_3 = g\, l/v^2$ $\equiv$ Fr^{-1} (Froude number)
$\Pi_4 = V/l^3$ = dimensionless displacement volume

The problem is thus completely defined by the Π set

$$\boxed{\{Ne, Re, Fr, V/l^3\}}$$

While maintaining geometrical similarity (V/l^3 = const), experiments on the scale of μ = 1:100 should be carried out to obtain Ne. However, the same Re and Fr cannot be set *simultaneously*, because when using the same liquid (water) the requirement Re = idem demands $v \times l$ to be constant, whereas Fr = idem demands v^2/l to be constant.

Supposing that the speed, v, of the model (index M) is specified by maintaining the same Fr value as in the full-scale (index H) application,

$$\text{Fr} = \text{idem}: (v^2/l)_M = (v^2/l)_H \quad \rightarrow \quad v_M = v_H\, \mu^{1/2}$$

then the same Re value in the model experiment must be attained by means of appropriate kinematic viscosity, ν:

$$\text{Re} = \text{idem}: (v\, l/\nu)_M = (v\, l/\nu)_H \quad \rightarrow \quad \nu_M = \nu_H\, \mu^{3/2}$$

For the model liquid at μ = 1:100 it therefore follows that $\nu_M = 10^{-3}\,\nu_H$. However, no fluid satisfies the condition $\nu = 10^{-3}\,\nu_{water}$!

If the scale-up factor were not necessarily that small (μ = 1:100!) and models were not so expensive to build, experiments could be run with models of various sizes in water at the same value of Fr, and the results extrapolated to $\text{Ne}(\text{Re}_H)$. In view of the powerful model reduction and the resulting extreme differences in the Reynolds number

$$\mu = 1\,;\, 0.1\,;\, 0.01 \quad \rightarrow \quad \text{Re}_M/\text{Re}_H = 1\,;\, 3.2\times10^{-2}\,;\, 1\times10^{-3}$$

extrapolation appears a risky undertaking, particularly when the cost of the motor used in the full-scale application is considered.

Naturally, the above results of dimensional analysis and their consequences were not known to the ship builders of the 19th century. Since the time of *Rankine*, the total drag resistance of the ship has been divided into three parts: the surface friction, the stern vortex and the bow wave. However, the concept of *Newton*ian mechanical similarity, known at that time, stated only that for mechanically similar processes the forces vary as $F \sim \rho\, l^2\, v^2$; scale up was not considered for assessing the effect of gravity.

Froude observed that the resistance due to the stern vortex was relatively small compared to the other two resistances and he decided to combine it with the bow wave resistance to obtain the *form drag* F_f. As a result of careful investigations and theoretical considerations, he realized that the wave formation of the ship could be simulated by using scale models. He arrived at the *law of appropriate velocities*:

> *"The wave formations at the ship and the model are (geometrically) similar, if the velocities are in the ratio of the square root of the linear dimensions".*

He also found that for similar wave formations, the hull drag (*friction drag F_r*) behaves not as $F_r \sim v^2 l^2 \rho$, but as $F_r \sim v^{1.825} A \rho$ (A - surface area), and he developed computational methods for scaling down models and ships by length and the type of the wetted surface. Thus, he was in a position to calculate the form drag F_f from the total drag after subtracting the predictable friction drag. He found:

> *"If we adhere to the law of the appropriate velocities in scaling-up the ship, the form resistances will correspond to the cubes of their dimensions (that is to say, their displacement volumes)"* [5][14].

In summary:

1. $F_{total} - F_r = F_f$
2. If $v^2 \sim l$, than $F_f \sim l^3$.

Dividing this functional dependence (2) by $\rho\, l^2 v^2$ in order to transform F_f into the Newton number of the form drag, requires that

$$\boxed{\frac{F_f}{\rho\, l^2 v^2} \sim \frac{l^3}{\rho\, l^2 v^2} = \frac{l}{\rho\, v^2} = \text{const.}}$$ This means:

$$\boxed{Ne_f = \text{idem} \quad \text{at} \quad Fr = \text{idem}} \quad \text{with} \quad Fr \equiv v^2/l\, g.$$

In order to verify these experimental results, the corvette "Greyhound" was towed by the corvette "Active" under the command of *Froude*, and the drag force in the tow rope was measured. The observed deviations from the predictions of the model were in the range of only 7 – 10% [17].

---

14) Reference [5] includes the minutes of the session of The Institution of Naval Archtects in London of April 7, 1870. At this session *W.Froude* presented and defended the results of his modeling with great steadfastness and conviction; this represents the sidereal hour of the theory of models.

M.Weber [A 1] points out that Froude´s procedure is not entirely correct and can never provide real proof, because complete similarity between the model and its full-scale counterpart cannot be achieved. The described procedure can therefore represent nothing more than an exellent approximation of reality[15]. He continues by saying: "The fact that *Froude* was able to achieve his goal with such a large measure of success despite all the difficulties, lies in his ingenuity which enabled him to itemize and to assess all the practical and theoretical details of drag resistance and finally to trace a clear picture of this intricate phenomenon." Froude´s performance cannot be judged too highly, especially in view of the measuring techniques available to him at that time!

Pawlowski[16] discussed an interesting alternative experimental approach to this scale-up problem. He also started by splitting the drag resistance into friction, depending only on Re, and a bow-wave resistance, depending only on Fr:

$$Ne_H = f_1(Re_H) + f_2(Fr_H).$$

However, he proposed a different strategy from that of *Froude*. In the *first* experiment, measurements are made with the model ship in water at $Fr_1 = Fr_H$, consequently $Re_1 = Re_H\,\mu^{-3/2}$, i.e., the measurement is carried out at a correct value of Fr and a false value of Re. As a result, a value of Ne_1 is obtained from the relationship:

$$Ne_1 = f_1(Re_1) + f_2(Fr_H).$$

Two additional experiments are carried out, not with the model ship, but with a totally immersed form (Fig.5) whose shape is given by reflecting the immersed portion of a ship´s hull at the water line (at V/l^3 = idem). In these experiments the Froude number is irrelevant; the friction corresponding to the surface area of the model must be divided by 2.

---

15) *M.Weber* is absolutely right: *Froude*´s approach is an excellent example of partial similarity!

16) This proposal was made by *J. Pawlowski* in his seminar on the Theory of Similarity in 1967.

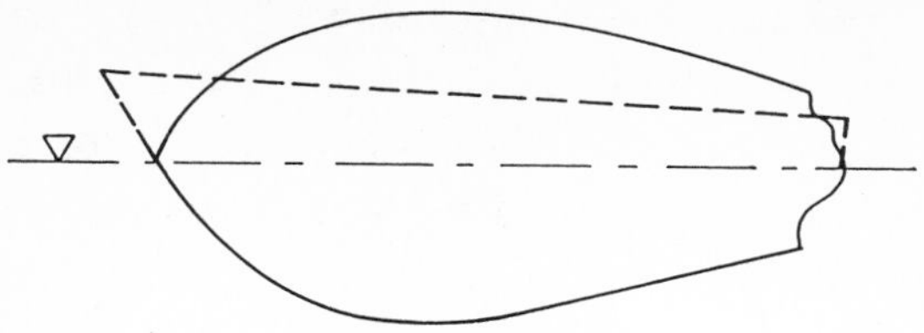

Figure 5: Sketch of the completelly submersed streamlined body

The measurements in water are carried out at Re_1 and Re_H, thereby obtaining

$Ne_2 = f_1(Re_H)$ und $Ne_3 = f_1(Re_1)$

The desired Ne_H can now be calculated:

$$\boxed{Ne_H = f_1(Re_H) + f_2(Fr_H) = Ne_1 - Ne_3 + Ne_2} \; .$$

The above two alternatives for calculation of the resistance of a ship illustrate an application of the method of partial similarity in which the process is divided into parts that can be investigated independently.

Example 2: *Mixing power per unit volume as a criterion for scale-up of mixing vessels*

Scaling up stirred tanks by using the criterion of a constant impeller power per unit volume, P/V, represents an important example of scale-up under conditions of partial similarity in chemical reaction engineering. It is obvious that the complicated fluid-mechanical processes which govern the mass and heat transfer in stirred tanks cannot generally be described adequately with the help of such a simple criterion.

The scale-up criterion P/V is only adequate in gas/liquid contacting and in liquid/liquid dispersion processes, when the impeller power is as uniformly dissipated as possible in the tank volume (micro-scale mixing, isotropic turbulence). See Example B4.

The most important mixing operation, i.e. homogenization of liquid mixtures, depends on the scale of the convective bulk transport (macromixing). Measurements resulted in the relationship $n\theta = f(Re)$, which depends heavily on the type of the stirrer and the geometry of the vessel, see [B 5] and Example A1, part 3.

Convective bulk transport is also extremely important in the suspension of solids in a stirred tank (which is also responsible for the flow pattern at the tank bottom). P/V cannot be used as a scale-up criterion in this process either. Measurements have shown that the minimum rotational speed n_{crit} of the stirrer which is necessary for the suspension (whirling-up) of particles in the turbulent regime is given by the appropriate Froude number: $Fr_{crit} \equiv n^2_{crit}\, d\, \rho/(g\, \Delta\rho)$.

What is the connection between P/V and the Froude number, the latter being the scale-up criterion? The answer is as follows:

Because the Froude number is the scale-up rule here, we start from $Fr \sim n^2 d$ (assuming the same material system) and express it in terms of

$$P/V \sim n^3 d^2$$

[$P \sim n^3 d^5$ (turbulent region and $V \sim d^3$; see Example A 1)]:

$$Fr \sim n^2 d \sim (n^2 d)^{3/2} = (P/V)\, d^{-1/2} = \text{idem}.$$

From this we obtain (the scale factor being defined as $\mu = d_M/d_H$):

$$[(P/V)\, d^{-1/2}]_H = [(P/V)\, d^{-1/2}]_M$$

$$\boxed{Fr = \text{idem} \rightarrow (P/V)_H = (P/V)_M\, \mu^{-1/2}}$$

Example 3: *Superficial velocity as a criterion for scale-up of bubble columns*

Bubble columns are often designed on the basis of the superficial velocity $v \equiv q/A$ (q - gas throughput, A - cross-sectional area of the column). Many authors have found that the gas/liquid mass transfer in bubble columns is indeed governed by this quantity (k_L - liquid side mass transfer coefficient;

a - interfacial area per unit volume):

$k_La \sim v \quad \rightarrow \quad k_La/v = \text{const.}$

This interdependence is only understandable when one considers that the volume-related mass transfer coefficient is defined by $k_La = G/V\Delta c$ and the superficial velocity v by $v = q/A$, as well the fact that volume $V = H \times A$ (H - height of the column):

$$\frac{k_La}{v} = \frac{G}{V\,\Delta c}\,\frac{A}{q} = \frac{G}{H\,q\,\Delta c} = \text{const.}$$

In words, the gas absorption rate $G[MT^{-1}]$ is proportional to the liquid height H, the gas throughput q, and the concentration difference Δc [B 8].

However, even for homogenization of the liquid content in the column by rising gas bubbles, the intensively formulated quantity v is not the only relevant parameter because, analogously to the same mixing operation in stirred tank (Example 2), liquid bulk transport (back-mixing) must take place over the entire height of the liquid column.

Experiments [B 9] performed on different scales gave the following expression for the mixing time θ in a bubble column of a given geometry:

$$\theta (g/D)^{1/2} \sim Fr^{-1/4}$$

where $Fr \equiv v^2/Dg$ and D is the column diameter. It follows that

$$\theta \sim v^{-1/2} D^{3/4} \qquad \text{or} \qquad \theta = \text{idem} \rightarrow v^{-1/2} D^{3/4} = \text{idem}$$

This leads to the conclusion that

$$\boxed{v_H = v_M\, \mu^{-1,5}}$$

Thus, in a bubble column geometrically scaled-up by a factor of $\mu = 10$, the same mixing time as in the model will be only obtained when the superfical velocity is increased by a factor of $10^{1,5} = 32$. Hence, v is not a scale-up criterion here.

Examples 2 and 3 show emphatically that *a particular scale-up criterion that is valid in a given type of apparatus for a particular process, is not necessarily applicable to other processes occurring in the same device.*

4 Treatment of variable physical properties by dimensional analysis

When using dimensional analysis to tackle engineering problems, it is generally assumed that the physical properties of the material system remain unaltered in the course of the process. Relationships such as the heat transfer characteristic of a technical device (e.g., vessel, pipe):
Nu = f (Re, Pr),
see example B1, are valid for any material system with Newtonian viscosity and for any constant process temperature, i.e. for *constant* physical properties.

However, constancy of physical properties cannot be assumed in every physical process. A temperature *field* may well generate a viscosity *field* or even a density *field* in the material system treated. In non-Newtonian (structurally viscous or elastoviscous) liquids, a shear rate can also produce a viscosity *field*.

Although most physical properties (e.g., viscosity, density, heat conductivity and capacity, surface tension) must be regarded as variable, it is particularly the value of viscosity that can be varied by many orders of magnitude under certain process conditions. In the following, dimensional analysis will be applied solely to describe the variability of this one physical property. However, the approach can be adapted for any other physical property.

4.1 Dimensionless representation of the material function

Similar behavior of a certain physical property common to different material systems can only be visualized by dimensionless representation of the material function of that property. It is furthermore desirable to formulate this function as uniformly as possible.

This can be achieved by the "standard representation" of the material function [9] in which a standardizing transformation of the material function $\eta(T)$ is defined such that the expression produced

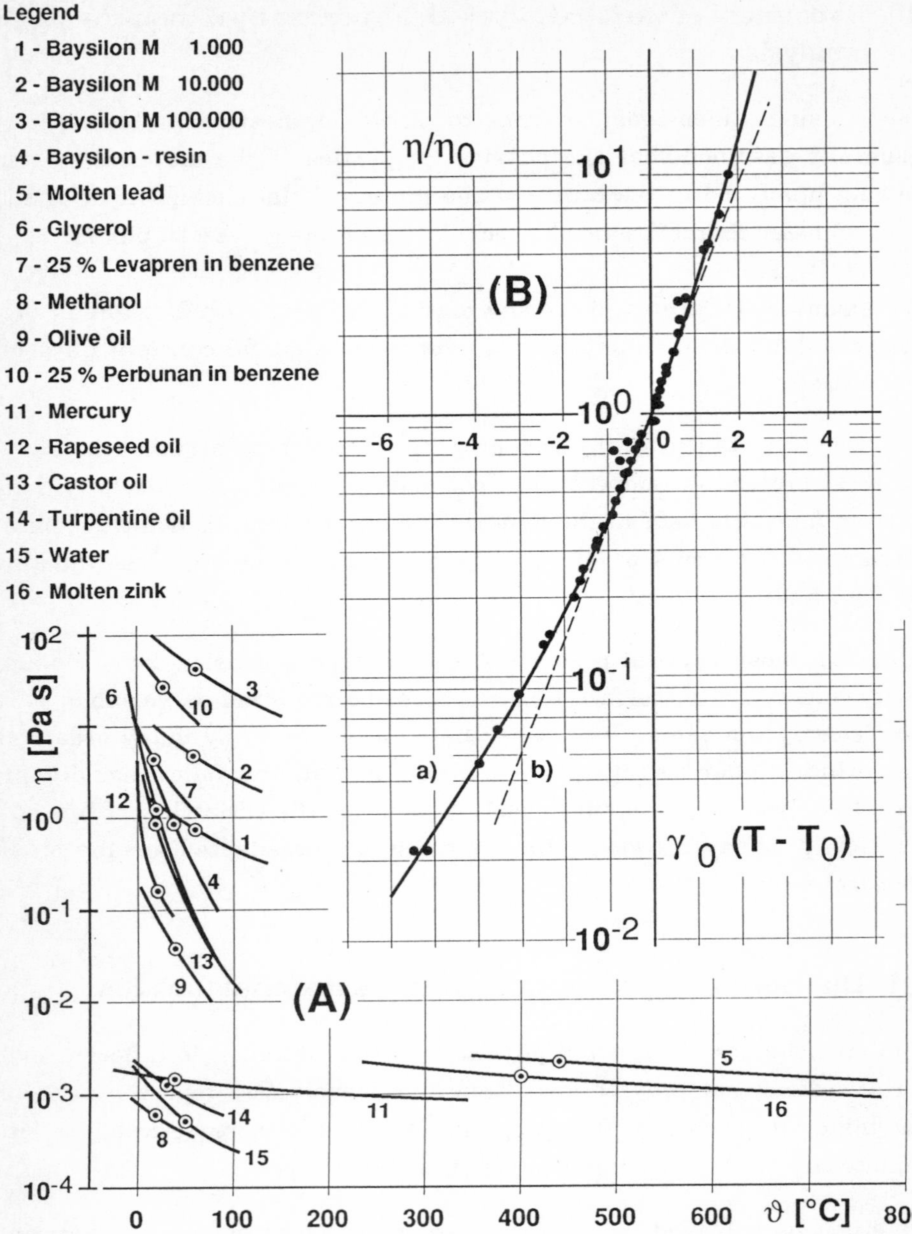

<u>Fig.6</u>: Temperature dependency of the viscosity (A) and its standard representation (B) which leads to the reference-invariant approximation of the material function [A 13]

$$\boxed{\eta/\eta_o = \phi\,\{-\,\gamma_o\,(T - T_o)\}} \qquad \text{(a)}$$

meets the requirement:
$\phi\,(0) = \phi'(0) = 1,$

where $\gamma_o \equiv (\frac{1}{\eta}\frac{\delta\eta}{\delta T})_{T_o}$ - temperature coefficient of the viscosity and

$\eta_o \equiv \eta\,(T_o)$. T_o is *any* reference temperature.

Figs 6A and 6B, taken from [A 13], depict the dramatic effect of this *standard transformation.*

Engineers prefer the representation

$$\boxed{\eta/\eta_o = \exp\,\{-\,\gamma_o(T - T_o)\}} \qquad \text{(b)}$$

which is a special case of (a) although this is not the best possible approximation (see the dotted line in Fig. 6 B). A slightly better approximation of the material function $\eta(T)$ is provided by the well known *Arrhenius* relationship[17] (cf. Fig. 1.3.3 in [A 13]):

$$\boxed{\frac{\eta}{\eta_o} = \exp\left\{-\frac{E_o}{RT_o}\left(\frac{T_o}{T} - 1\right)\right\}} \qquad \text{(c),}$$

In order to describe the *process* by dimensional analysis, it is advisable to formulate the reference temperature T_o in a process-related manner, using a characteristic, possibly mean process temperature as reference. A class of functions exists, however, which are independent (invariant) of the reference point T_o. The solid line in Fig. 6 B shows a function of this type. (Note: Functions (a) and (b) are special cases of invariant functions [9].)

---

17) This representation was first used by *Swante Arrhenius* (1859 - 1927) to describe the temperature dependence of the reaction rate constant $k = k_\infty \exp(E/RT)$.

4.2 The Π set for variable physical properties

The type of dimensionless representation of the material function affects the (extended) Π set within which the process relationship is formulated. When the standard representation is used, the relevance list must include the reference viscosity η_0 instead of η and incorporate two additional parameters γ_0, T_0. This leads to *two* additional dimensionless numbers in the process characteristic. As regards the above-mentioned heat transfer characteristic of a device,
$Nu = f(Re, Pr)$,
it now follows that

$$\boxed{Nu = f(Re_0, Pr_0, \gamma_0 \Delta T, \Delta T/T_0)} \ ,$$

where the index $_0$ in Re and Pr denotes that these two dimensionless numbers are to be formed with η_0 (which is the numerical value of η at T_0).

If we take into account the fact that the standard transformation of the material function can be expressed invariantly with regard to the reference temperature T_0 (Fig. 6 B), then the relevance list is extended by only *one* additional parameter γ_0. This, in turn, leads to only *one* additional dimensionless number. For the above problem it now follows that

$$\boxed{Nu = f(Re_0, Pr_0, \gamma_0 \Delta T)} \ .$$

In process characteristics for heat transfer, the temperature dependence of the *density* is normally not taken into account by the dimensionless number $\beta \Delta T$ (β is the temperature coefficient of the density, defined in a way similar to that used for γ). Instead, it is usual to employ the *Grashof* number Gr thus taking into account the fact that the density differences can only be effective in the presence of gravity:

$$Gr \equiv \beta \Delta T \, Ga = \beta \Delta T \, Re^2 \, Fr^{-1} = \beta \Delta T \, g \, l^3 \, \nu^{-2}.$$

For more in-depth information on sections 4.1 and 4.2 , see [A 9, Chap. 3]; [A 13], [21], [23].

4.3 Treatment of non-Newtonian liquids by dimensional analysis

The main characteristic of Newtonian liquids is that simple shear flow (e.g., Couette flow) generates shear stress τ which is proportional to the rate of shear D. The proportionality constant, the dynamic viscosity η, is the only material constant in the law of flow:

$$\boxed{\tau = \eta \boldsymbol{D}}$$

η depends only on pressure and temperature.

In the case of non-Newtonian liquids, η depends on D as well. These liquids can be classified in various categories of materials depending on their flow behaviour (c.f. DIN 1342/1&2 and 13342). The graphic representation of flow behavior using $D(\tau)$ is called a flow curve; when $\eta(D)$ or $\eta(\tau)$ is used, it is called the viscosity curve.

Figure 7 depicts the viscosity curve for a mixture of steam engine cylinder oil and ca. 7% aluminum stearate. The asymptote *a* corresponds to the so-called yield point $\tau_0 = 50$ N/m^2; at a high rate of shear $\eta_\infty = 9.7$ Pa s = const. is reached (asymptote *b*). Figure 8 shows a dimensionless representation of the viscosity curves of various chemical and biological polymers (rheological material functions).

Pawlowski ([A 9, Chap.5] and [22]) points out that the rheological properties of many non-Newtonian liquids can be described by material parameters whose dimensional matrix has a rank of two. These physical properties can be usefully linked to produce two dimensional constants

a characteristic viscosity constant H and

a characteristic time constant θ

and possibly a set of dimensionless material numbers Π_{rheol}. The relevance list is thus formed by

H, θ, Π_{rheol}.

(In the example of Fig.7 $H = b$ und $\theta = H/\tau_0 = 0.194$ s)

Changing from a Newtonian to a non-Newtonian liquid, the above interrelation has the following consequences with regard to the complete Π set:

a) every dimensionless number incorporating η must now be formulated with the dimensionally equivalent H (e.g., η_∞);
b) a single process-related number containing θ is added;
c) the pure material numbers are increased by Π_{rheol}.

We will demonstrate these rules on the heat transfer characteristic of a smooth, straight pipe. A stands for a temperature *in*dependent and B for a temperature dependent viscosity:

	Newtonian liquid	non-Newtonian liquid
A	Nu, Re, Pr	Nu, Re_H, Pr_H, $v\theta/L$
B	Nu, Re, Pr, $\gamma_o \Delta T$	Nu, Re_H, Pr_H, $v\theta/L$, $\gamma_H\,\Delta T$, γ_θ/γ_H

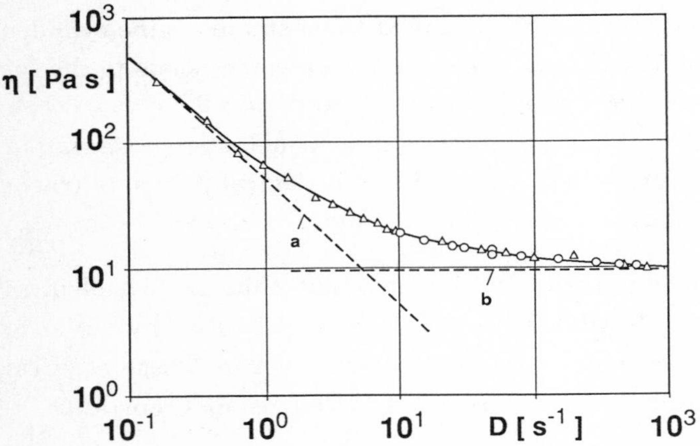

Fig.7: Viscosity curve for a mixture of steam engine cylinder oil and approx. 7% aluminum stearate (from [A 9])

In B $\gamma_H \equiv \delta \ln H/\delta T$ and $\gamma_\theta \equiv \delta \ln \theta/\delta T$ have to be added. ΔT denotes the temperature difference between the bulk of the liquid and the wall. The conventionally used expression η/η_w (index w - wall) and $\gamma \Delta T$ are related by:

$$\eta/\eta_w = \exp\,[-\gamma\,(T - T_w)].$$

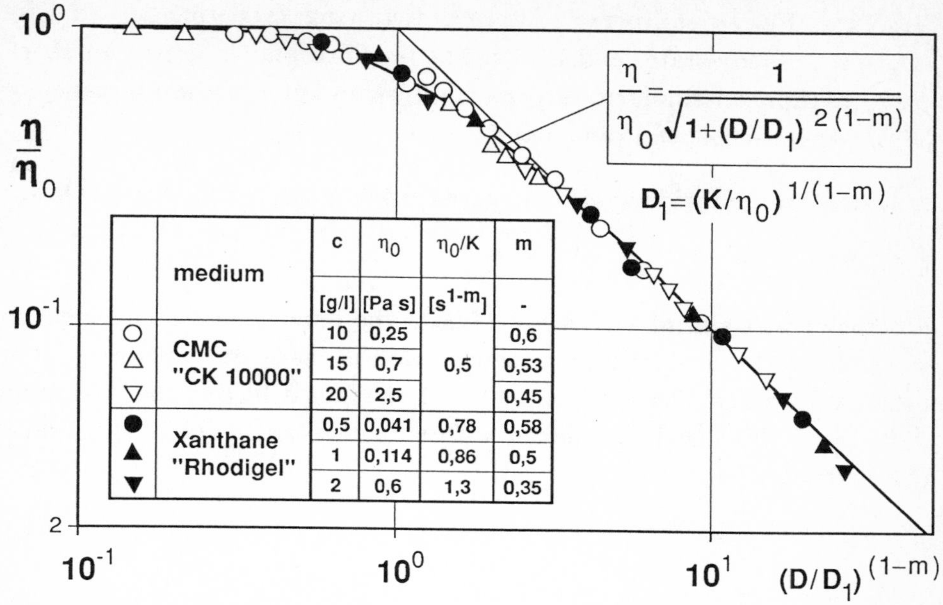

	medium	c [g/l]	η_0 [Pa s]	η_0/K [s^{1-m}]	m [-]
○	CMC "CK 10000"	10	0,25	0,5	0,6
△		15	0,7		0,53
▽		20	2,5		0,45
●	Xanthane "Rhodigel"	0,5	0,041	0,78	0,58
▲		1	0,114	0,86	0,5
▼		2	0,6	1,3	0,35

Fig.8: Dimensionless representation of viscosity curves of different chemical und biological polymers (from [23])

Since little is known about the rheological properties of material systems, model experiments must be carried out with the same substance as will be used in the full-scale plant. Since

$\Pi_{material}$ (here Pr_H) = idem and Π_{rheol} = idem

the process will then take place within a Π range with only one more dimensionless number ($v\,\theta/L$) than in the case of Newtonian liquids.

However, when scaling-up from the model to the full-scale plant, it is impossible to attain complete similarity using the same material system. In the above example, it is obvious that in keeping ρ, H, θ = idem it is impossible to ensure that $Re_H \equiv \rho\, v\, L/H$ and $v\,\theta/L$ also remain identical. It is therefore advisable to retain one and the same substance but to alter the scale of the model in the experiments, cf. Sect. 3.2 and examples A6 and

A7. Pure fluid-mechanical processes involving creeping (ρ being irrelevant), steady-state, and isothermal flow are an exception. In these cases mechanical similarity can be obtained in spite of the constancy of physical properties, cf. example A 10.

Non-Newtonian liquids with flow curves obeying the so-called power law

$$\tau = K \boldsymbol{D}^m.$$

are known as *Ostwald - de Waele Fluids*. The dimension of K depends on the numerical value of the exponent m and is therefore not a consistent physical quantity. To produce a material constant X having the dimension of viscosity, K must be combined with a characteristic velocity, v, and a characteristic length, l:

$$X \equiv K(v/l)^{m-1}.$$

With this sort of liquid, the extension of the Π set has the following consequences:

a) X replaces η in every dimensionless number in which it appears;
b) pure material numbers are extended by m (m belongs to Π_{rheol} !).

In the case of an Ostwald- de Waele fluid the heat transfer characteristic of a smooth straight pipe reads as follows:

$$Nu = f(Re_X, Pr_X, m),$$

where $Re_X = \rho\, v^{2-m}\, l^m\, X^{-1}$. (To learn more about the questionability of this Π space, which originates from the inconsistency of K, see [9] und [A 14].)

4.4 Treatment of viscoelastic liquids by dimensional analysis

Almost every biological solution of low viscosity (but also viscous biopolymers like xanthane, Fig. 8) and dilute solutions of long-chain polymers (carboxy-methyl cellulose CMC (Fig. 8) polyacrylamide, polyacrylnitrile, etc.) display not only viscous but also viscoelastic flow behavior. These liquids are capable of storing a part of the deformation energy elastically and reversibly. They evade mechanical stress by contracting like rubber bands. This behavior causes secondary flow which often runs contrary to the flow produced by mass forces (e.g. the liquid "climbs" the shaft of a stirrer, the so-called "Weissenberg effect").

Elastic behaviour of liquids is characterized mainly by the ratio of first differences in normal stress, N_1, to the shear stress, τ. This ratio, the Weissenberg number $Wi = N_1/\tau$, is usually represented as a function of the rate of shear $\boldsymbol{D}$.

Henzler [23] takes an approach similar to that of the Ostwald - de Waele law:

$$Wi = A\boldsymbol{D}^a \qquad \text{or preferably} \qquad Wi = A\boldsymbol{D}^a + B\boldsymbol{D}^b$$

To transform these material functions into dimensionless forms, a reference Weissenberg number Wi_0 is chosen. This leads to the generation of the characteristic rates of shear D_1 and D_2. It thus follows that:

$$\boxed{Wi/Wi_0 = (\boldsymbol{D}/\boldsymbol{D_1})^a + (\boldsymbol{D}/\boldsymbol{D_2})^b} \; .$$

In this case the following constituent parts of the dimensionless material function have to be incorporated in the relevance list:

$Wi_0, \boldsymbol{D_1}, \boldsymbol{D_2}$, a, b.

In the case of a viscoelastic liquid the material functions of both the viscous and the elastic behavior have to be considered. Assuming that the viscous behaviour of the liquid concerned is described as in Fig. 8 and the above Wi/Wi_0 function is valid for its elastic behaviour, the Π set must be modified as follows:

a) η_0 replaces η in every dimensionless number in which η appears.

b) The pure material-related numbers are extended by the following *six* purely rheological numbers:
$\Pi_{rheol} = Wi_0, \boldsymbol{D}/\boldsymbol{D_1}, \boldsymbol{D}/\boldsymbol{D_2}$, m, a, b.

For rheological characteristics of fermentation broth see [24].

APPENDIX

EXAMPLES OF PRACTICAL APPLICATION

FROM THE FIELD OF

CHEMICAL ENGINEERING

A Examples from the field of mechanical unit operations

Introductory remarks:

Fluid mechanics and mixing operations in various types of equipment, disintegration and agglomeration, mechanical separation processes, etc. are described by parameters, the dimensions of which consist of only three primary quantities: **M**ass, **L**ength and **T**ime. An isothermal process course is assumed: the physical properties of the material system treated are related to a *constant* process temperature. The process relationships obtained in this way are therefore valid for any constant, random process temperature to which the numerical values of the physical properties are related, as long as there is no departure from the scope of validity of the process characteristic verified by the tests.

As far as the field of mechanical unit operations is concerned, the scale-up can only present problems in the case of non-availability of model substances. (Best example: Drag resistance of a ship´s hull and *Froude´s* approach of this problem, see p. 42/47)

Example A 1:

Power consumption and mixing time for the homogenization of liquid mixtures. Design principles for stirrers and the determination of optimum conditions (minimum mixing work $P\theta$)

In order to design and dimension stirrers for the homogenization of liquid mixtures - and this is by far the most common task when it comes to stirring! - it is vital to know the *power characteristic* and the *mixing time characteristic* of the type of stirrer in question. If this information is available for various types of stirrers it is possible to determine both the best type of stirrer for the given mixing task and the *optimum operating conditions* for this particular type.

a) Power characteristic of a stirrer

The power consumption P of a given type of stirrer under given installation conditions depends on the following variables:
Geometric parameter: Stirrer diameter d
Physical properties: Density ρ and kinematic viscosity ν of the liquid
Process parameter: Rotational speed n of the stirrer

The relevance list therefore reads as follows:

$\{P; d; \rho, \nu; n\}$

The (appropriately assembled) dimensional matrix undergoes only one linear transformation to produce the two dimensionless numbers:

	ρ	d	n	P	ν
M	1	0	0	1	0
L	−3	1	0	2	2
T	0	0	−1	−3	−1

Z_1	1	0	0	1	0
Z_2	0	1	0	5	2
Z_3	0	0	1	3	1

$Z_1 = M$
$Z_2 = L + 3M$
$Z_1 = -T$

$\Pi_1 = P/(\rho\, d^5\, n^3) \equiv Ne$ (Ne - Newton number)
$\Pi_2 = \nu/(d^2\, n) \equiv Re^{-1}$ (Re - Reynolds number)

The experimentally determined *power characteristic* Ne(Re) of a blade stirrer of a given geometry under given installation conditions (see sketch in Fig. A1.3) is presented in Fig. A1.1. It follows from this that:

1) $Ne \sim Re^{-1}$ and thus $\boxed{NeRe \equiv P/(\eta\, n^2\, d^3) = const.}$ is valid in the range Re < 20. Density is irrelevant here - we are dealing with the *laminar* flow region.

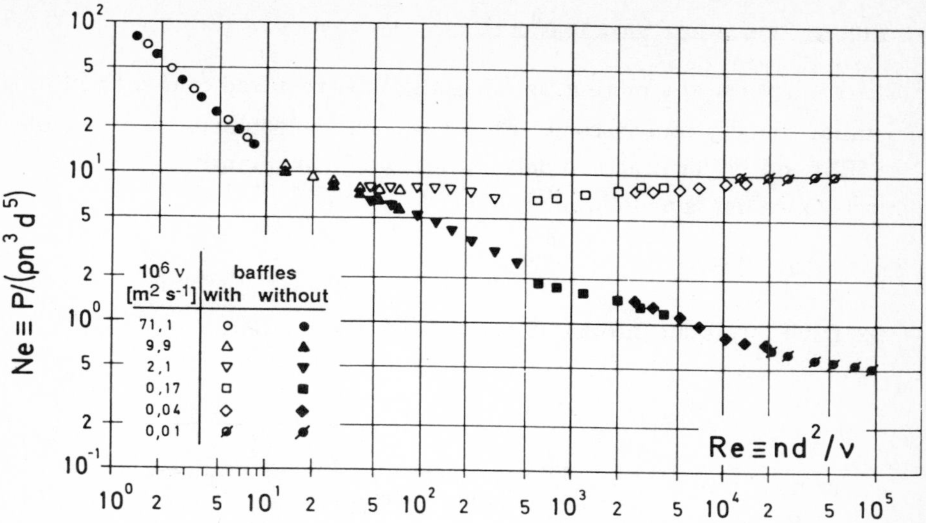

Fig. A1.1: Power characteristic of a blade stirrer of a given geometry under given installation conditions (see Fig. A1.3); from [B 5]

2) $\boxed{Ne \equiv P/(\rho\, n^3\, d^5) = \text{const.}}$ remains valid in the range Re > 50 (vessel with baffles) or Re > 5×10^4 (unbaffled vessel). In this case, viscosity is irrelevant - we are dealing with a *turbulent* flow region.

3) The influence of the baffles is, understandably, nil in the laminar flow region. However, it is extremely strong at Re > 5×10^4. The installation of baffles under otherwise unchanged operating conditions increases the stirrer power by a factor of 20 here!

4) When stirred in an unbaffled vessel, the fluid begins to circulate and a vortex is formed. The question whether gravitational acceleration g, and hence the Froude number $Fr \equiv n^2\, d/g$, plays a role under these circumstances can be safely answered in the negative on the basis of the test results shown in Fig. A1.1: For confirmation, one need only look at the points on the lower Ne(Re) curve where the *same* Re value was set for fluids with *different* viscosities. This was only possible by a proportional alteration of the rotational speed of the stirrer. Where Re = idem, Fr was clearly not idem, but this has no influence on Ne: g is therefore irrelevant!

b) Mixing time characteristic of a stirrer

For our purposes, the mixing time θ is the time required to mix two fluids of similar density and viscosity until they are molecularly homogeneous. (See [B5] and [18] for practical determination of this quantity). In this case, the relevance list is as follows:

$\{\theta; d; \rho, \nu, \mathfrak{D}; n\}$

where $\mathfrak{D}[L^2\,T^{-1}]$ is the diffusivity of one fluid in the other.

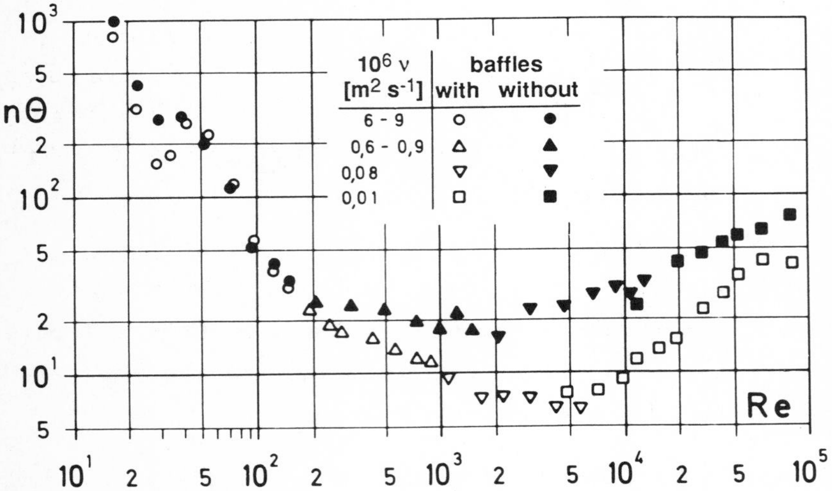

Fig.A1.2: Mixing time characteristic of a blade stirrer; see Fig.A1.3 for geometrical conditions

The dimensional analysis of this example is described in Section 2.4, p. 37. It was concluded that:

$\{n\theta; Re \equiv n\,d^2/\nu; Sc \equiv \nu/\mathfrak{D}\}$ resp. $\boxed{n\theta = f(Re, Sc)}$

represents the appropriate Π space here. The corresponding Π relationship, the *mixing time characteristic* of the blade stirrer for the geometric

[18] P.V. Dankwerts, Chem. Eng. Sci. **7** (1957), 116/117 ; J.W.Hiby, Fortschr. der Verfahrenstechnik **17** (1979) B, 137/155 (in English)

conditions given in Fig. A1.3, is shown in Fig.A1.2. The measured values demonstrate, in analogy to point 4 of the discussion of the power characteristic, that $\mathfrak{D}$, and consequently the Schmidt number, are not as relevant as assumed. (In the same system of materials - a mixture of water and cane syrup - at almost the same $\mathfrak{D}$ value, the Schmidt number is varied by the kinematic viscosity ν over decades.)

c) Optimum conditions for the homogenization of liquid mixtures

If the power and mixing time characteristics are known for a series of common stirrer types under favorable installation conditions, see [B 5], one can go on to consider optimum operating conditions by asking the question: Which type of stirrer operates within the requested mixing time θ with the lowest power consumption and hence the minimum mixing work ($P\theta$ = min) in a given system of materials and a given vessel (vessel diameter D)?

In answering this question, we need not (at least for the moment) consider the diameter of the stirrer or its speed of rotation; the relevance list is as follows:

$\{P, \theta; D; \nu, \rho\}$

When the dimensional matrix has been assembled as follows:

	ρ	D	ν	P	θ
M	1	0	0	1	0
L	−3	1	2	2	0
T	0	0	−1	−3	1
Z_1	1	0	0	1	0
Z_2	0	1	0	−1	2
Z_3	0	0	1	3	−1

$Z_1 = M$
$Z_2 = L + 3M + 2T$
$Z_1 = -T$

we are left with two dimensionless numbers:

$$\Pi_1 \equiv \frac{P\,D}{\rho\,\nu^3} = \frac{P\,D\,\rho^2}{\eta^3} \quad \text{and} \quad \Pi_2 \equiv \frac{\theta\,\nu}{D^2} = \frac{\theta\,\eta}{D^2\rho}$$

These two dimensionless numbers can be formed from the known numerical values of Ne, nθ and Re with the help of D/d. The following interrelations exist:

$\Pi_1 \equiv \mathrm{Ne\ Re\ D/d}$ and $\Pi_2 \equiv \mathrm{n\theta\ Re^{-1}\ (D/d)^2}$

Fig. A1.3 shows this relationship for those stirrer types exhibiting the lowest Π_1 values within a specific range of the dimensionless number Π_2, i.e. the stirrers requiring the least power in this range.

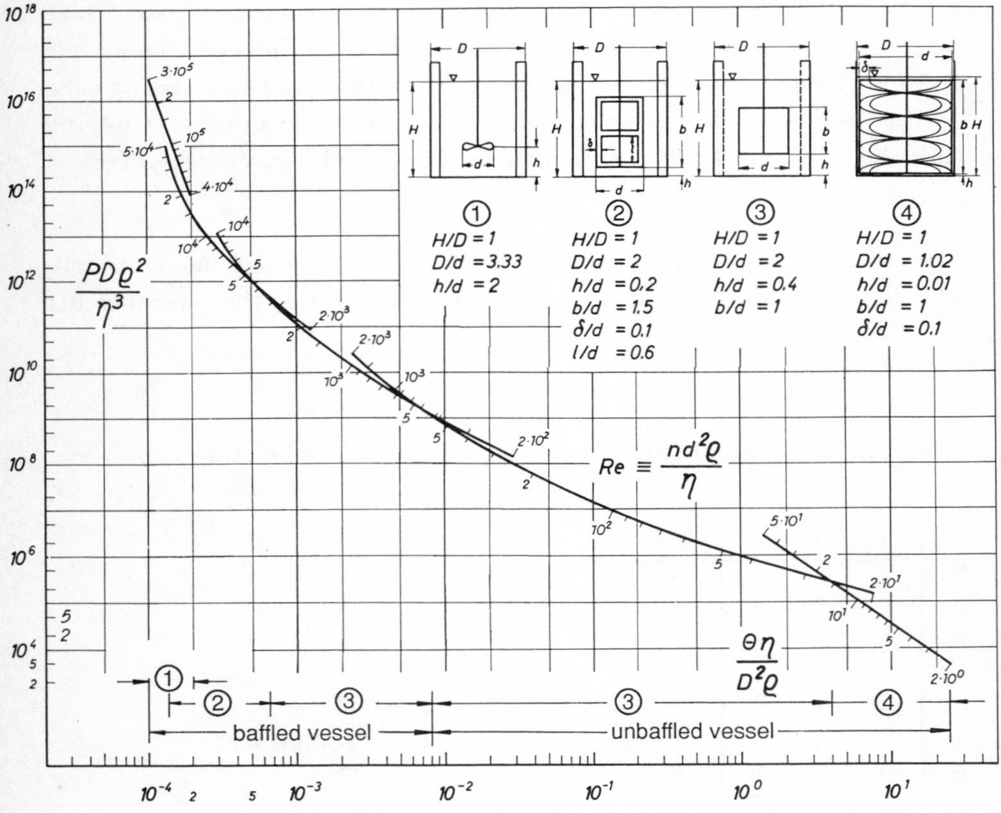

Fig.A1.3: Working sheet for the determination of optimum working conditions on the homogenization of liquid mixtures in mixing vessels. From [B 5].

This graph is extremely easy to use. The physical properties of the material system, the diameter of the vessel (D) and the desired mixing time (θ) are all known and this is enough to generate the dimensionless number Π_2. The curve $\Pi_1 = f(\Pi_2)$ in Fig. A1.3 then provides the following information:

a) The *stirrer type* and baffling conditions can be read off the abscissa. The diameter of the stirrer can be determined from data on stirrer geometry in the sketch.

b) The numerical value of Π_1 can be read off at the intersection of the Π_2 value with the curve. The *power consumption* P can then be calculated from this.

c) The numerical value of Re can be read off the Re scale at the same intersection. This, in turn, makes it possible to determine the *rotational speed* of the stirrer.

Example A 2:

Power consumption in the case of gas/liquid contacting. Design principles for stirrers and model experiments for scale-up

Gas/liquid contacting is frequently encountered in chemical reaction and bioprocess engineering. For reactions in gas/liquid systems (oxidation, hydrogenation, chlorination, etc.) and aerobic fermentation processes (including biological waste water treatment), the gaseous reaction partner must first be absorbed by the liquid. In order to increase its absorption rate the gas must be dispersed into fine bubbles in the liquid. A fast rotating stirrer (e.g. a turbine stirrer), to which the gas is supplied from below, is normally used for this purpose (see sketch in Fig. A 2.1).

For a given geometry of the set-up, the relevance list for this problem contains the power consumption P as the required target quantity, the stirrer diameter d as the characteristic length and a number of physical properties of the liquid and the gas (the latter are marked with an apostrophe): Densities ρ and ρ', kinematic viscosities ν and ν', surface tension σ and an unknown number of still unknown physical properties S_i which describe the coalescence behaviour of finely dispersed gas bubbles and, indirectly, their hold-up in the liquid. The process parameters are the rotational speed n of the stirrer and the gas throughput q, which can be adjusted independently, as well as the gravitational acceleration g which is implicitly relevant because of the large density difference. (We should actually have written $g\Delta\rho$ here – see Sect. 2.1.4, p. 27 – but, since

$\Delta\rho = \rho_{liquid} - \rho_{gas} \approx \rho_{liquid}$, the dimensionless number would contain $g\Delta\rho/\rho_{liquid} \approx g\rho_{liquid}/\rho_{liquid} = g$. The relevance list is therefore:

$\{P; d; \rho, \nu, \sigma, S_i, \rho', \nu'; n, q, g\}$

In this case the relevance list contains at least 11 variables - more than twice the number for the power consumption in the homogeneous liquid system (Example A1a)!

Before employing the dimensional analysis for generation of the dimensionless numbers, it is worthwhile anticipating obvious numbers such as ρ'/ρ and ν'/ν. S_i are physical properties of unknown dimension and number. Therefore they cannot be included in the dimensional matrix. However, this is no great problem since, with the known relevant physical properties ρ, ν, σ, we will always be able to transform S_i to the dimensionless numbers S_i^*. The above relevance list can therefore be reduced to

$\{P; d; \rho, \nu, \sigma; n, q, g\}$ for anticipated $\rho'/\rho, \nu'/\nu, S_i^*$

Here, the simplest dimensional matrix is also the best one because it leads directly to the common, named dimensionless numbers.

	ρ	d	n	P	ν	σ	q	g
M	1	0	0	1	0	1	0	0
L	−3	1	0	2	2	0	3	1
T	0	0	−1	−3	−1	−2	−1	−2

Z_1	1	0	0	1	0	1	0	0	$Z_1 = M$
Z_2	0	1	0	5	2	3	3	1	$Z_2 = L + 3M$
Z_3	0	0	1	3	1	2	1	2	$Z_3 = -T$

$\Pi_1 = P/(\rho\, d^5\, n^3) \equiv Ne$ (Ne - Newton number)

$\Pi_2 = \nu/(d^2\, n) \equiv Re^{-1}$ (Re - Reynolds number)

$\Pi_3 = \sigma/(\rho\, d^3\, n^2) \equiv We^{-1}$ (We - Weber number)

$\Pi_4 = q/(d^3\, n) \equiv Q$ (Q - gas throughput number)

$\Pi_5 = g/(d\, n^2) \equiv Fr^{-1}$ (Fr - Froude number)

Taking the anticipated numbers into account, it follows that:

$$\boxed{Ne = f\,(Re, We, Q, Fr, \rho'/\rho, \nu'/\nu, S_i^*)}$$

However, we can now see that the important process parameter, the rotational speed of the stirrer n, is present in all numbers: Their numerical value is changed with each change in speed. This is not advantageous for the planning and evaluation of experiments. Our aim is therefore to transform Re and We, which contain the physical properties ν and σ, to dimensionless material numbers through combination with other numbers.

Firstly we form two combinations of dimensionless numbers which do not contain n:

$Re^2/Fr \equiv g\, d^3\, \nu^{-2} \equiv Ga$ (Galilei number)

$We/Fr \equiv \rho\, d^2\, g\, \sigma^{-1}$

Then we have to combine these two new numbers in such a way as to eliminate d:

$$\{(Re^2/Fr)^{2/3}\,(Fr/We)\} \equiv (Re^4\, Fr)^{1/3}\, We^{-1} \equiv \sigma(\rho^3\, \nu^4\, g)^{-1/3} \equiv \sigma^*$$

By this means it is possible to transform the We number into a pure material number σ^*, the numerical value of which is dependent only on the material system. In contrast, the only advantage of the Ga number over Re is that it does not contain the rotational speed. Let us stay with Re and base our considerations on the following Π space:

$$\boxed{Ne = f\,(Q, Fr, Re;\, \sigma^*, \rho'/\rho, \nu'/\nu, S_i^*)}$$

In this case, the Π space consists of a target number (Ne), three process numbers (Q, Fr, Re) and a series of pure material numbers).

The first question we must ask is: Are laboratory tests performed in *one single piece* of laboratory apparatus - i.e. on one single scale - capable of providing binding information on the decisive process number (or combination of numbers)? Although we can change Fr by means of the rotational speed of the stirrer, Q by means of the gas throughput and Re (or Ga) by means of the liquid viscosity independent of each other, we must accept the fact that a change in viscosity will alter not only Re but also the numerical values of the material numbers σ^*, ν'/ν and, very probably, S_i^*.

In contrast to the gas density ρ' [19), an influence of the gas viscosity ν' on the stirrer power is not to be expected (ν'/ν = irrelevant). Preliminary tests with methanol/water mixtures showed [B6] that σ does not influence the stirrer power either, σ^* = irrelevant. Furthermore, measurements revealed that the coalescence behaviour of the material system is not affected if aqueous glycerol or cane syrup mixtures are used to increase viscosity. This means that the influence of S_i^* on Ne cannot be distinctive. These results alone give us the right to perform model experiments in *one single piece of apparatus* in order to elaborate the process relationship. The following Π space is used as the basis

$$\boxed{\mathrm{Ne} = f(Q, \mathrm{Fr}, \mathrm{Re})}$$

when *one* gas (air) is used ($\rho'/\rho \approx$ const).
The results of these model experiments are described in detail in [B6]. For our consideration based on the theory of similarity, it is sufficient to present only the main result here. This states that, in the industrially interesting range ($\mathrm{Re} \geq 10^4$ und $\mathrm{Fr} \geq 0.65$), Ne is dependent only on Q; see Fig.A2.1.

Knowledge of this power characteristic, the analytical expression for which is

$$\mathrm{Ne} = 1.5 + (0.5\, Q^{0.075} + 1600\, Q^{2.6})^{-1}$$

19) The buoyancy of hydrogen bubbles differs strongly from that of air bubbles. Therefore it must be expected that the influence of ρ' on the gas hold-up will be relatively strong.

can be used to reliably design a stirrer drive for the performance of reactions in the gas/liquid system (e.g. oxidation with O_2 or air, fermentation etc.) as long as the physical, geometric and process-related (Re and Fr range) boundary conditions comply with those of the model measurement.

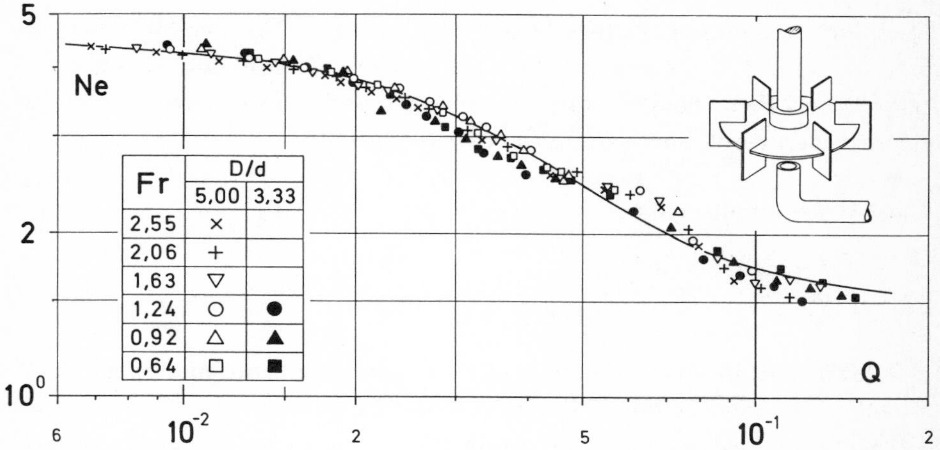

Fig.A 2.1: Power characteristic of a turbine stirrer under industrially interesting conditions (Re ≥ 10^4 und Fr ≥ 0.65). From [B6]

At this point we should ask ourselves how we would perform scale-ups if we did not know anything about the above functional relationship! Let us therefore try to determine the stirrer power of a given stirrer for a specified large fermenter (e.g., V = 100 m^3; H/D = 3; D = 3.5 m) on the basis of model measurements where the physical properties of the system and the gas throughput of interest are known. With a freely selectable rotational speed of the stirrer, we can presuppose that, apart from the numerical value of Nu, the numerical values of all other dimensionless numbers are known. (Of course we do not know anything about the coalescence phenomena - corresponding to the state of knowledge existing about 15 years ago.) Our considerations are therefore based on the following relationship:

$$\boxed{Ne = f(Q, Fr, Re; \sigma^*, \rho'/\rho, \nu'/\nu)}$$

Naturally we will perform the model experiments (e.g., on a scale of μ = 1:10, i.e. V = 0.1m^3 $\rightarrow$ D = 0.35 m) with the industrially interesting material system; in this way at least the numerical values of the three material numbers remain constant. However, this means that Re (resp. Ga) *and* Fr can no longer be adjusted *independently* because only the rotational speed of the stirrer is available for their realization.

Therefore we can only realize *partial similarity* in the model: We can either set Q and Re = idem or Q and Fr = idem. We will opt for the second case (Q and Fr = idem) because we expect g and hence Fr to be more important then ν and hence Re in the case of aeration.

From the scale-up rules

$$Fr \sim n^2 d = \text{idem} \rightarrow n_M = n_H \mu^{-1/2} \qquad \text{and}$$
$$Re \sim n d^2 = \text{idem} \rightarrow n_M = n_H \mu^{-2} \qquad (\mu = d_M/d_H)$$

it follows that, in order to maintain Fr = idem, the rotational speed of the stirrer in the model (index M) is only faster by a factor of $\mu^{-1/2}$ than in the full-scale application (index H) while the condition Re = idem requires a stirring rate which is faster by a factor of μ^{-2}. This means that if we carry out the model measurement with Fr = idem, we are running the risk of the flow condition shifting substantially towards the laminar region with respect to Re: That is to say, if we perform the model experiments, which are characterized with

$$Fr = \text{idem:} \qquad n_M = n_H \mu^{-1/2} \qquad \text{and} \qquad d_M = d_H \mu$$

the following is valid for the Re number under these conditions:

$$Re_M \sim n_M d^2_M = n_H \mu^{-1/2} (d_H \mu)^2 \sim Re_H \mu^{3/2}$$

In our example (μ = 1:10), the Re number would only be approximately 1/30th of the value obtained in the full-scale application! In view of this fact, the following approach would seem to be sensible, <u>Fig. A 2.2</u>:

1. The first measurement point is determined at Re_{M1} and Fr, Q = idem (filled circle in the figure).

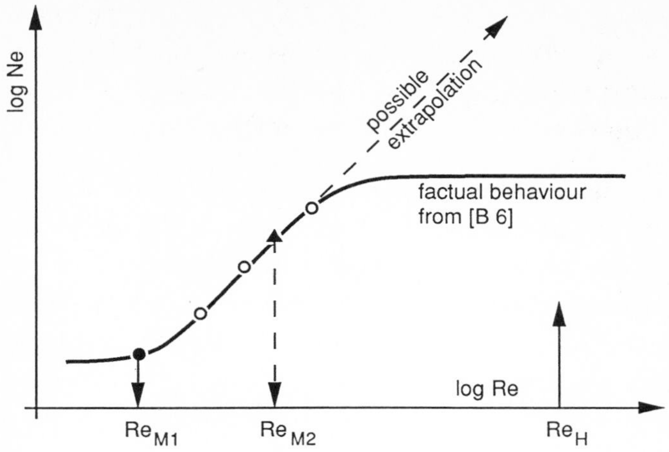

Fig. A 2.2: Presentation of the Ne(Re) curve for Q, Fr = idem kept constant. Illustration of model experiments. For explanation see text.

2. In the course of further measurements at Fr, Q = idem, the viscosity of the experimental liquid is reduced stepwise to raise Re towards Re_H; compare the three hollow circles. The smaller the selected model scale, the greater is the danger of this approach also finally leading to false extrapolation to $Ne(Re_H)$: We do not *know* that Re is no longer relevant at $Re \geq 10^4$!

3. Since the material numbers are changed by the approach described in point 2, we will perform another model experiment on a larger scale at Re_{M2} just to be on the safe side, see filled triangle. Although we may confirm preceding measurement results, this in no way reduces the risk of *extrapolation* to Re_H, as shown by comparison with the actual state of affairs (curve in Fig.A 2.2 according to [B6]).

This example is not intended as a deterrent against measurements under the conditions of partial similarity in general since these – when performed with care – can frequently provide valuable information. The purpose of this consideration is merely to show that there can be no substitute for *complete* information about a technical matter in the right Π space.

Example A 3:

Power consumption and gas throughput in self-aspirating hollow stirrers. Optimum conditions for P/q(Gas) = min and an answer to the question whether this type of stirrer is suitable for technical applications

As a result of their form, hollow stirrers utilize the suction generated behind their edges to suck in gas from the head space above the liquid. As "rotating ejectors", they are stirrers and conveying devices at one and the same time and are therefore particularly suitable for laboratory use (especially in high-pressure autoclaves) because they offer intensive gas/liquid contacting via internal gas recycling without a separate gas conveying device.

a) Power characteristic of the hollow stirrer

When compared to conventional stirrers (A1a, p.64), the gas density ρ' and gravitational acceleration g must be added to the relevance list because these two parameters also govern the gas hold-up and consequently the effective density of the system. Since g can only act in combination with $\Delta\rho = \rho - \rho'$, see Section 2.1.4, there is only one additional parameter in this case: $g\Delta\rho$. This becomes $g\Delta\rho/\rho \cong g$ in the corresponding dimensionless number, see relevance list for A2, p.70. We will therefore introduce g at once. Furthermore, it is also important to remember that the gas throughput is not a freely selectable parameter in the case of hollow stirrers and, as a result, cannot be included in the relevance list. It follows that:

$\{ P; d; \nu, \rho; g, n \}$

	ρ	d	n	P	ν	g
M	1	0	0	1	0	0
L	−3	1	0	2	2	1
T	0	0	−1	−3	−1	−2

Z_1	1	0	0	1	0	1	$Z_1 = M$
Z_2	0	1	0	5	2	1	$Z_2 = L + 3M$
Z_3	0	0	1	3	1	2	$Z_3 = -T$

$\Pi_1 \equiv P/(\rho\, d^5 n^3) \equiv Ne$ (Ne – Newton number)
$\Pi_2 \equiv \nu/(d^2 n) \equiv Re^{-1}$ (Re – Reynolds number)
$\Pi_3 \equiv g/(d^2 n) \equiv Fr^{-1}$ (Fr – Froude number)

The Π framework then reads:

$\{ Ne, Fr, Re \}$

To establish the Π relationship of a so-called triangular hollow stirrer (see sketch in Fig.A3.1) measurements [B10] were performed under given installation conditions on the scale $\mu = 1 : 2 : 3 : 4 : 5$. Fig. A 3.1 shows that the Reynolds number is irrelevant for stirring in water ($Re > 10^4$) and that, as a result, the *power characteristic* is given by Ne(Fr) alone.

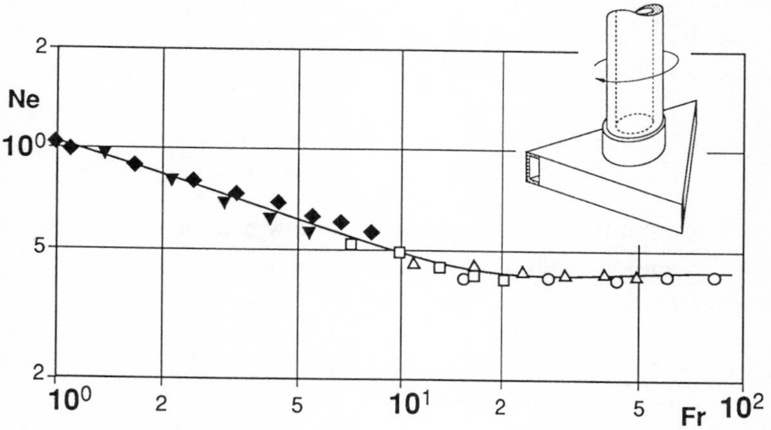

Fig.A3.1: Power characteristic of the triangular stirrer. Installation conditions: H/D = 1; D/d = 3.3; h/d = 1 (H - liquid height; D - vessel diameter; h - ground clearance of the stirrer). A legend is given in Fig. A 3.2. From [B10].

b) Gas throughput characteristic of a hollow stirrer

If the target quantity power P is replaced by the self-aspirated gas throughput q in the above relevance list and in the dimensional matrix, the Π set is:

{ Q, Fr, Re }	with	$Q \equiv q/(nd^3)$ - gas throughput number.

The associated Π relationship is illustrated in Fig.A3.2.

The measurements verify that, in this case too, the Reynolds number does not exert any influence at $Re > 10^4$.

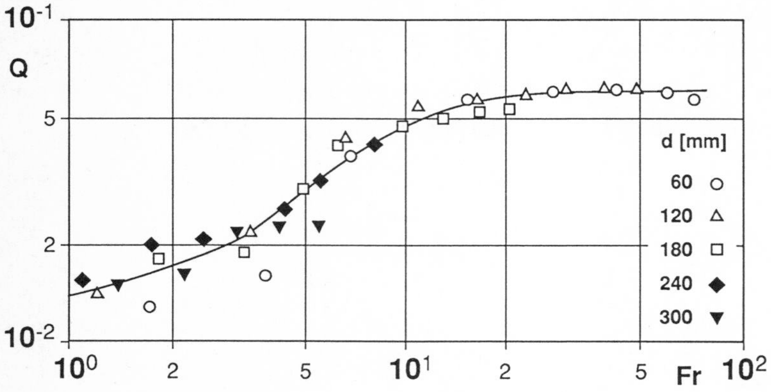

Fig.A 3.2: Gas throughput characteristic of a hollow stirrer (same stirrer type and same installation conditions as in Fig. A 3.1). From [B10].

c) Optimum conditions to achieve the desired gas throughput q with the minimum power consumption (P/q = min)

In order to identify these conditions, we have to form a new number from the numbers Ne, Fr and Q. It is a dimensionless expression of the quotient P/q and - like its components - must be a function of the Froude number:

$$\boxed{Ne\, Fr\, Q^{-1} \equiv P/(q\, d\, \rho g) = f(Fr)}$$

If the numerical values of this dimensionless number are plotted as a function of the Fr number, it becomes apparent that they do not exhibit a minimum in the industrially interesting range of Fr = 2 - 10 but rather have a value of approx. $10^2 \approx$ const which does not allow any optimization. The following numerical equation is therefore valid for triangular hollow stirrers and for the material system water/air:

$$\boxed{P/q\ [kWh/m^3] = 0.28\ d\ [m]}$$

While a stirrer with a diameter of d = 1 m requires a specific power of P/q = 280 Wh/m^3, a stirrer which is 10 times smaller needs only 28 Wh/m^3!

The question posed in the title as to whether hollow stirrers are suitable for industrial application can be answered in the negative simply from the structure of the new dimensionless number:

$$Fr = idem \quad \rightarrow \quad P/(q\ d\ \rho g) \sim P/(q\ d) = idem$$

$$\boxed{(P/q)_H = (P/q)_M\ \mu^{-1}}$$

If the model apparatus is scaled up on a scale of μ = 1:10 and operated at Fr = idem, 10 times more power will be required per gas throughput than in the laboratory.

Conclusion: Hollow stirrers are not suitable for the self-aspiration of large quantities of gas on an industrial scale! In this case, the gas throughput and stirrer power must be decoupled from each other, for instance using a fast rotating turbine stirrer to which separately compressed gas is supplied from below; see Example A2.

Example A 4:
Mixing of solids in drums with axially operating paddle mixer

In the final instance, the mixing of solids (e.g., powders) can only lead to a stochastically homogeneous mixture. We can therefore use the theory of random processes to describe this mixing operation. In the present example from [B11] we will concentrate on a mixing device in which the position of the particles is adequately given by the x coordinate. Furthermore, we will assume that the mixing operation can be described as a stochastic process without "aftereffects". This means that only the actual condition is important and not its history. The temporal course of this so-called *Markov* process can be described with the second *Kolmogoroff* equation. In the case of a mixing process without selective convectional flows (requirement: $\Delta\rho \approx 0$ and $\delta_i \approx 0$; compare the studies on segregation phenomena in beds of spherical particles[20]), the solution of *Fick*'s diffusion equation gives a cosine function for the local concentration distribution, the amplitude of which decreases exponentially with the dimensionless time $\theta\, \mathfrak{D}/(\pi^2\, l^2)$, see Fig.A 4.1. (The variation coefficient v is defined here as the standard deviation of the relative concentration deviation.)

Let us now consider this process using the principle of dimensional analysis. We have the following parameters:

v - variation coefficient as a measure for the mixing quality; D, l - diameter and length of the drum; d - diameter of the mixing device; δ - mean particle diameter; ϕ - fill degree of the drum; $\mathfrak{D}$, ρ - diffusivity and density of the particles, n - rotational speed of the mixer; θ - mixing time; $g\rho$ - solid gravity.

The relevance list contains 11 parameters

$\{\, v;\, D, l, d, \delta, \phi;\, \mathfrak{D}, \rho;\, n, \theta, g\rho \,\}$

After the exclusion of the dimensionless quantities v and ϕ and the obvious geometric dimensionless numbers l/D, d/D and δ/D, the following 3 dimensionless numbers are obtained:

	ρ	D	n	θ	$\mathfrak{D}$	$g\rho$
M	1	0	0	0	0	1
L	−3	1	0	0	2	−2
T	0	0	−1	1	−1	−2

Z_1	1	0	0	0	0	1	$Z_1 =$ M
Z_2	0	1	0	0	2	1	$Z_2 =$ L + 3M
Z_3	0	0	1	−1	1	2	$Z_3 = -$T

20) M. Ullrich, Chem.-Ing.-Tech. **41** (1969) 16, 903/907

θ n		mixing time number
$\mathcal{D}/D^2n$	$\equiv Bo^{-1}$	Bo – Bodenstein number
$g\rho/\rho Dn^2$	$\equiv Fr^{-1}$	Fr – Froude number

The complete Π set reads:

$$\{ v, L/D, d/D, d/D, \phi, \theta n, Bo, Fr\}$$

To obtain the rotational speed of the drum in only one number (the process number Fr), we combine the other two accordingly with Fr and obtain: $\theta\, \mathcal{D}/D^2$ und $g\, D^3/\mathcal{D}^2$.

The experimental results presented in Fig.A 4.1 were obtained in one single model (D = 0.19 m) with different lengths (l/D = 1; 1.5; 2; 2.5). The geometric and physical numbers d/D, δ/D, ϕ and $g\, D^3/\mathcal{D}^2$ remained unchanged as did Fr because of the constant rotational speed of the drum

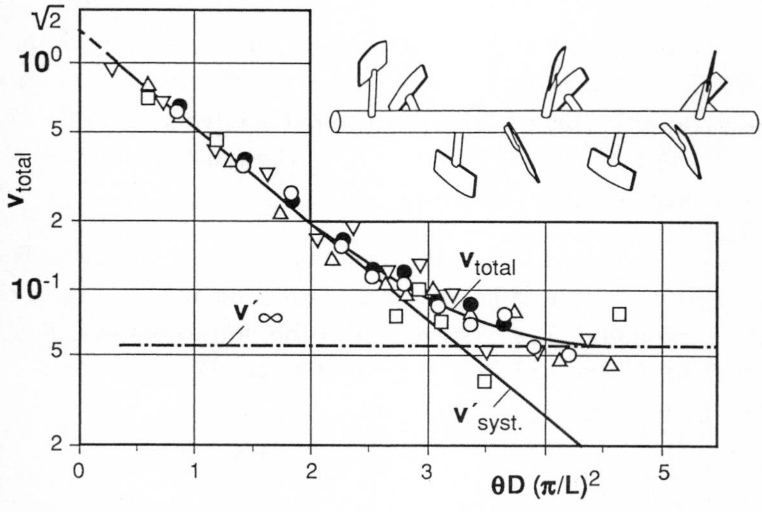

Fig.A 4.1: Mixing quality as a function of the dimensionless mixing time for different l/D ratios. Copper and nickel particles of δ = 300 - 400 µm, fill degree of the drum ϕ = 35%, Froude number of the drum Fr = 0.019. From [B 11].

$n = 50\ min^{-1}$. As a result, the measurements can only be depicted in the Π space

$$\{ v, \theta \mathcal{D}/D^2, l/D \}$$

whereby d/D, δ/D, ϕ, $g\ D^3/\mathcal{D}^2$, Fr = idem.

The result of these measurements is

$$v = f(\theta\ \mathcal{D}/l^2)$$

In other words, the mixing time required to attain a certain mixing quality increases with the square of the drum length. In order to reduce the mixing time, the component to be mixed in will have to be added in the middle of the drum or simultaneously at several positions .

Example A. 5

Gas hold-up in bubble columns and its dependence on geometric, physical and process-related parameters

Bubble columns are important pieces of apparatus for the absorption of gases in liquids and, consequently, for the execution of chemical reactions in the gas/liquid system. In this context, the attainable interface (= sum of the surfaces of all gas bubbles) is of most interest because it affects the mass flow in a directly proportional manner.

If a gas throughput q is input into a bubble column with the diameter D and the liquid height H, the liquid height rises by the amount occupied by the gas bubbles in the liquid. The gas fraction in the liquid, the so-called gas hold-up h can be determined from the liquid height H_b of the gassed and the non-gassed liquid (see Fig. A 5.1).

$$\frac{V_{gas}}{V_{liquid}} = \frac{H_b - H}{H} \equiv h$$

In the course of extensive measurements on bubble columns with different dimensions [B 12], the gas hold-up h proved to be directly proportional to the volume-related mass transfer coefficient k_La (see

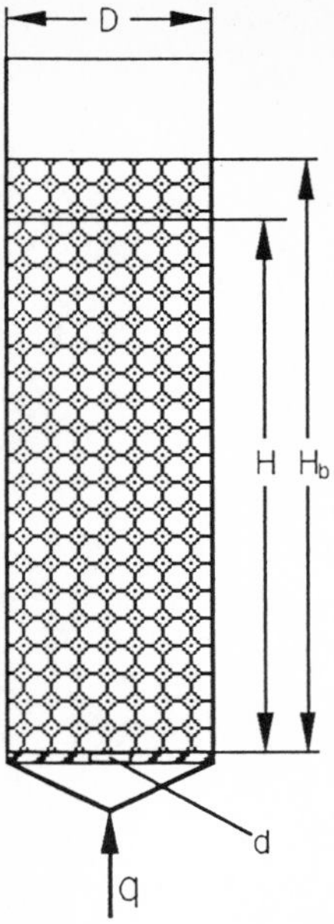

Fig. A 5.1: Sketch of a bubble column

Example B4 with reference to k_La). For this reason, h will be the target number in the following considerations.

Bubble columns with a single-hole plate as gas distributor (see sketch) were used in the investigations. D, H and d (= hole diameter) therefore describe their *geometry* in full. The densities (ρ and ρ') and the viscosities (ν and ν') of both phases ($'$-gas) and the surface tension σ must be taken into account as *physical properties*. The *process parameters* are the gas throughput q and, on account of the extreme differences in density, the gravity difference $g\Delta\rho = g\,(\rho - \rho')$.

The complete relevance list is therefore:

$\{\, h;\, D, H, d;\, \rho, \rho', \nu, \nu', \sigma;\, q, g\Delta\rho \,\}$

If we exclude the target number h and the trivial numbers H/D, d/D, ρ'/ρ and ν'/ν, the remaining relevance list is

$\{\, D;\, \rho, \nu, \sigma;\, q, g\Delta\rho \,\}$

from which the following 3 dimensionless numbers result:

	ρ	D	ν	σ	q	gΔρ
M	1	0	0	1	0	1
L	−3	1	2	0	3	−2
T	0	0	−1	−2	−1	−2

Z_1	1	0	0	1	0	1	$Z_1 = M$
Z_2	0	1	2	3	3	1	$Z_2 = L + 3M$
Z_3	0	0	1	2	1	2	$Z_3 = -T$

Z'_1	1	0	0	1	0	1	$Z'_1 = M$
Z'_2	0	1	0	−1	1	−3	$Z'_2 = L + 3M + 2T$
Z'_3	0	0	1	2	1	2	$Z'_3 = -T$

$$\Pi_1 = \frac{\sigma D}{\rho \nu^2} \equiv \frac{Re^2}{We} \qquad \Pi_2 = \frac{q}{D \nu} \equiv Re \qquad \Pi_3 = \frac{g \Delta\rho\, D^3}{\rho \nu^2} \equiv Ar \equiv \frac{Re^2}{Fr}$$

This dimensional analysis produces two dimensionless numbers (Π_1 and Π_3) which, apart from the column diameter, contain only physical properties, and the Reynolds number Π_2 as the process number (because it contains q).

However, this is not particularly satisfactory because we must expect the hydrodynamics of a bubble column to be substantially governed by gΔρ, as a result of which the process number must contain gΔρ. We must therefore recombine Π_2 and Π_3 to obtain the modified Froude number Fr* as the probably *true process number*:

$$\Pi_2^2\, \Pi_3^{-1} \equiv \frac{q^2 \rho}{D^5\, g \Delta\rho} \equiv Fr^*$$

The complete Π set is now:

{ h; H/D, d/D; ρ′/ρ, ν′/ν, Re²/Fr, Ar, Fr*}

Comprehensive measurements [B 12] were performed to verify this Π space and to evaluate the process characteristic:

a) The relationship h = f(Fr*) was examined using a bubble column of given geometry; water was used as liquid and the physical properties of the gas were varied over a wide range. Air, nitrogen and nitrogen/hydrogen mixtures were used. The result in Fig.A 5.2 demonstrates that Fr* takes full account of the influence of both densities and that ν'/ν is obviously irrelevant (see footnote 19) in Example A 2 on page 72)! The process relationship is:

$$h = 15.0\ Fr^{*0.35}$$

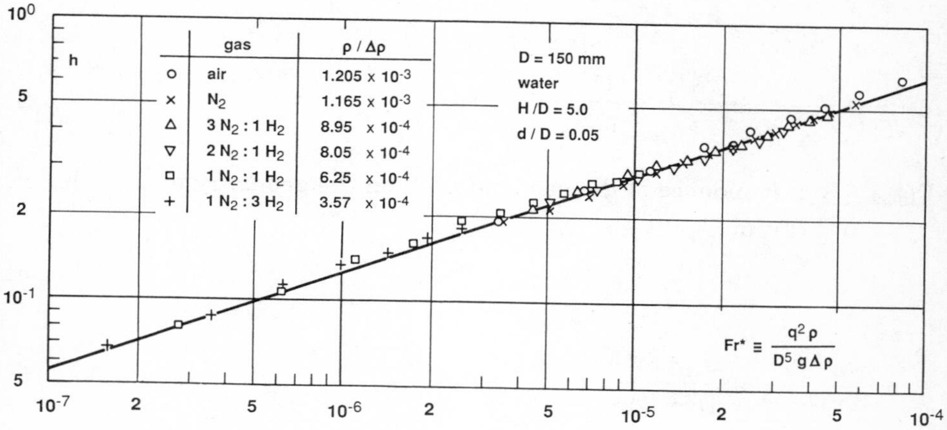

Fig.A 5.2: The relationship h(Fr*) confirms that the gas hold-up h in a bubble column is dependent on the process number Fr*. From [A 12].

b) Further measurements were carried out with one single material system (water/air) in bubble columns of different geometries (Fig. A 5.2). The extended process relationship now reads:

$$h = 16.7\ Fr^{*0.35}\ (H/D)^{-0.25}\ (d/D)^{-0.125}$$

c) The influence of the physical properties of the liquid phase was investigated in one single model bubble column using water and in addition 12 different pure organic liquids, the physical characteristics of which vary substantially, see Table 2 in [B12]. The result of these measurements is presented in Fig.A 5.4. The dimensionless numbers had to be combined as follows to correlate the measured values:

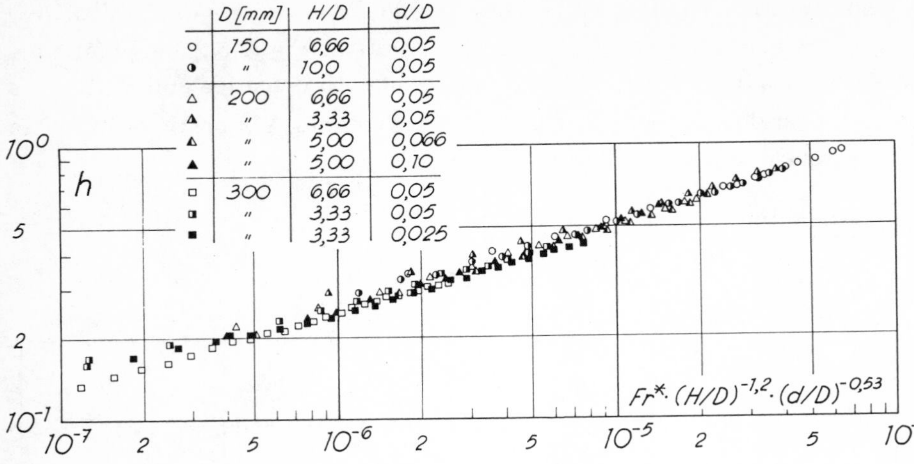

Fig.A 5.3: Dependence of the gas hold-up h on the geometry of the column. From [A 12].

$$B \equiv \frac{\rho/\rho' \ Fr^*}{Re\ We^{0.5}} \equiv \frac{\nu\ (\sigma\,\rho)^{0.5}}{D^{2.5}\ g\Delta\rho}$$

The disadvantage of the resulting number combination B is that it is not a pure material number, but still contains the column diameter. For safe scale-up, the correctness of this correlation will therefore have to be checked on columns of different diameters and with simultaneous alteration of the material system!

The correlation from Fig.A 5.4 may give the impression that it will always be possible to depict an n-digit Π space two-dimensionally using the analytical evaluation. This is not necessarily the case. The more complicated the physical facts are, the more incomplete the analytical description and depiction will be. Indeed it is quite easy to imagine situations where this will not even be possible. One example is the pressure drop characteristic curves of the straight, smooth pipe in Figures 3 and 4 (p. 34/35), the analytical reproduction of which is likely to cause considerable problems!

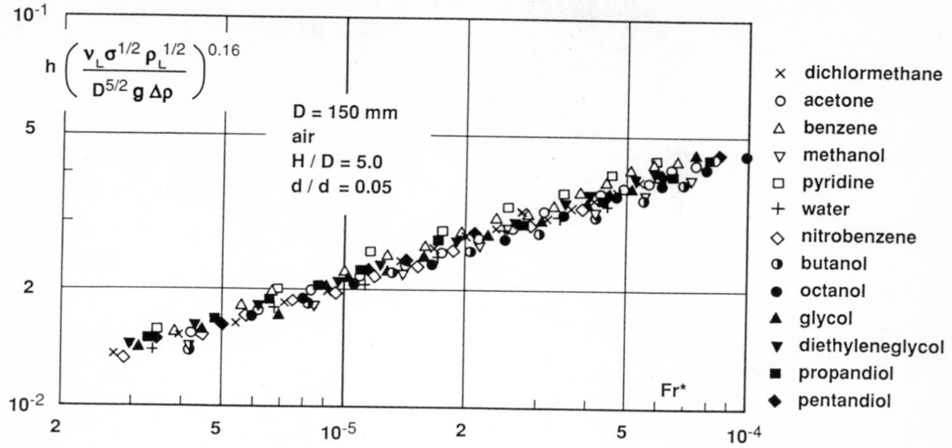

Fig.A 5.4: Influence of the physical properties of both phases on the gas hold-up h in a bubble column of given geometry.

Example A 6:

Description of the flotation process with the aid of two intermediate quantities

In the past few years, flotation techniques have also been used increasingly in waste water treatment and process water recycling[21]. They are also suitable for removing activated sludge from biologically treated waste water, especially if it tends to float on the surface (e.g., as in the case of waste water from the food processing industry).

A distinction is made between *induced air* flotation and *dissolved air* flotation whereby, in the latter case, the waste water to be treated is practically saturated with air at about 6 bar in advance. *Degasifying flotation* of biologically treated effluent is a special variant of this technique. It is used if the waste water is biologically treated in high towers ("Tower Biology") and therefore contains a lot of CO_2 which, after it has been submitted to fast degasification, flotates the activated sludge.

---

21) M. Zlokarnik, Chem.-Ing.-Tech. **53** (1981) 8, 600/606 or Ger. Chem. Eng. **5** (1982) 2, 109/115 and Kem. Ind.(YU - Zagreb) **34** (1985) 1, 1/6 (in German)

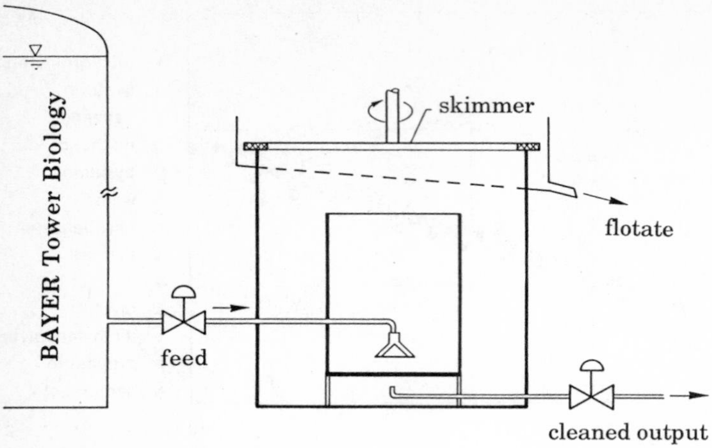

Vertical flow flotation cell with separate **aeration** and **separation** zones for dissolved gas flotation

Fig. A 6.1: Sketch of the flotation cell exibiting vertical flow with spatially separated flotation and tranquillizing chambers for degasifying flotation

A possible set-up for this flotation technique is presented in Fig.A 6.1. It consists of a flotation cell which executes a vertical flow direction and is equipped with a funnel-shaped nozzle, in which the liquid is suddenly exposed to suction. As a result, the gas dissolved in the liquid is degasified spontaneously, floating the solid matter [B 2].

Let us now consider this process using dimensional analysis. As far as the *target* quantity is concerned, it should be remembered that flotation is a depletion process taking place according to the time law of the 1st order which, in the case of batch-wise performance, is described by the flotation rate constant k [T^{-1}]:

$$\ln(\varphi_t/\varphi_o) = -k\,t$$

Naturally, dissolved air flotation and its variant degasifying flotation are only possible if the flotation cell operates continuously. In this case the *solids discharge A* can be selected as the target quantity:

$$A = 1 - (\varphi_{out}/\varphi_{in})$$

If we could assume that the liquid flow in the cell were ideally backmixed (this is not the case in the cell shown in Fig. A 6.1!), we would be able to link this depletion process, which fulfils the time law of the 1st order, with the residence time distribution characteristic of a completely backmixed unit to obtain:

$$A = 1 - (\varphi_{out}/\varphi_{in}) = 1 - (1 + k\,\tau)^{-1}$$

with: A solids discharge; k flotation rate constant; τ = mean residence time of the throughput in the unit.

What is the (dimensionless) *target* quantity solids discharge A dependent on? The cell *geometry* is characterized by the cell diameter D. In contrast, the *physical properties* which permit and describe the flotation process are not so readily available. We know that the degree of hydrophobicity (wettability) of the particle surface is described by the contact angle Θ. Furthermore, we know that the electrical charge of the particles depends greatly on the pH. Very fine particles can only be made to float on the surface after they have been cross-linked with the aid of a polyelectrolyte (its concentration being c_{PE}) to form flakes. Further influencing parameters will be the gravity difference $g\Delta\rho$ between the solid particle and the liquid, the solids content φ in the liquid and, of course, the average gas bubble size (the parameter S_i describing coalescence). (Our present knowledge of the flotation process does not allow us to name all relevant physical properties, let alone measure them!)
In contrast, it is easy to list the *process* parameters:
Liquid feed q_{in}, which leaves the flotation cell divided into flotate output $q_{flotate}$ and "clear" output: $q_{out} = q_{in} - q_{flotate}$;
releasable gas content of the liquid feed $q_{gas}/q_{in} = Hy\,\Delta p/\rho_{gas}$;
gravitational acceleration g. (Hy - Henry constant of gas solubility)

The relevance list, which is certainly incomplete with regard to the physical properties, is:

$$\{ A; D; \Theta, pH, c_{PE}, g\Delta\rho, \varphi, S_i, q_{in}, q_{out}, q_{gas}/q_{in} \}$$

This relevance list comprising at least 11 parameters can be streamlined significantly by introducing two intermediate quantities:

1. The *superficial velocity w* in the annulus of the flotation cell (Fig. A6.1):

$w = 4\, q_{out} / \pi\, (D^2 - D'^2)$ and

2.the *rising velocity* v of the flakes:

$v = f\, (\Theta, pH, c_{PE}, g\Delta\rho, \varphi, S_i, q_{gas}/q_{in})$

The introduction of the rising velocity of the flakes v as an intermediate variable should allow *all* relevant physical properties to be taken into account; however it presupposes separate measurement of this variable in suitable equipment, see [B 2].

The complete relevance list is then:

$\{A;\, w, v, q_{in}, q_{out}\}$

and provides the following 3-parametric Π set:

$\boxed{\{A,\, w/v,\, q_{in}/q_{out}\}}$

In order to verify this relationship, tests were performed on various days in a flotation cell such as that illustrated in Fig.A 6.1 with D = 1.6 m and q_{in}/q_{out} = const. with the discharge from a biological sewage treatment plant (BAYER "Tower Biology"® in Leverkusen-Bürrig; liquid height 26.5 m; CO_2 content of the off-gas ca. 8 - 10 % by volume). The parameter w/v was varied by changing the rising velocity of the flakes through the addition of a polyelectrolyte, see table in Fig. A 6.2.

Fig. A 6.2 shows the result of these measurements. It confirms that it was possible to satisfactorily describe the solids discharge A with the process number w/v for uniform (on a daily average) chemical composition of the biologically treated mixed waste water from a large chemical company. On the other hand, it indicates - at least in the case of chemical waste water - the existence of waste water components whose influence on the floatability of the sewage sludge is not taken into account by the rising velocity of the flakes v alone. Therefore the expectations for the above relevance list and the resulting 3-parametric Π space are not completely fulfilled.

Flotation is a very simple and effective separation technique for both the mechanical and biological waste water treatment and for the recycling of

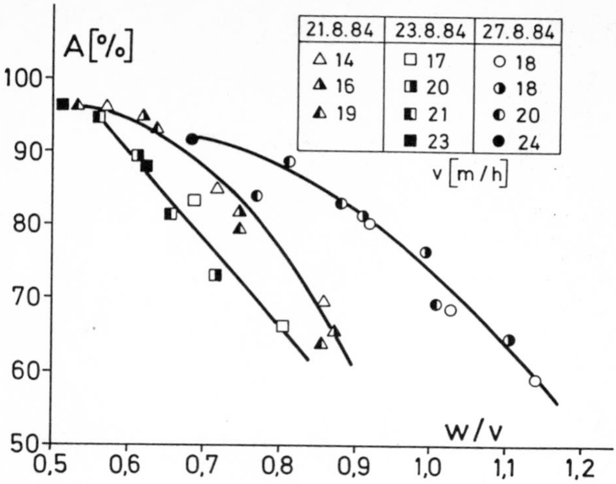

Fig. A 6.2: Relationship A (w/v) for removal of biosludge by means of flotation from the discharge from a industrial tower biology facility. D = 1,6 m; q_{in}/q_{aut} = const. Taken from [B 2].

process water. Its importance will certainly increase in the future. Therefore it will be worth preparing data for the dimensioning and scale-up of flotation devices. The contribution made by Example A6 should be understood as an incentive in this direction.

Example A 7:

Preparation of design and scale-up data for mechanical foam breakers without knowledge of the physical properties of the foam

Foam control is becoming increasingly important with the introduction of new separation processes based on foam formation (flotation, stripping, foaming out) and with the technical realization of new microbiological processes. The metabolic processes frequently produce surface-active substances ("surfactants") responsible for the generation of stable foams in the fermentation broth which is generally made up of 3 to 4 phases.

Foam can be broken thermally, chemically or mechanically but only the last possibility is suitable for biological techniques, see [B13]. This example deals with a mechanical foam breaker as shown in Fig. A 7.1 which uses

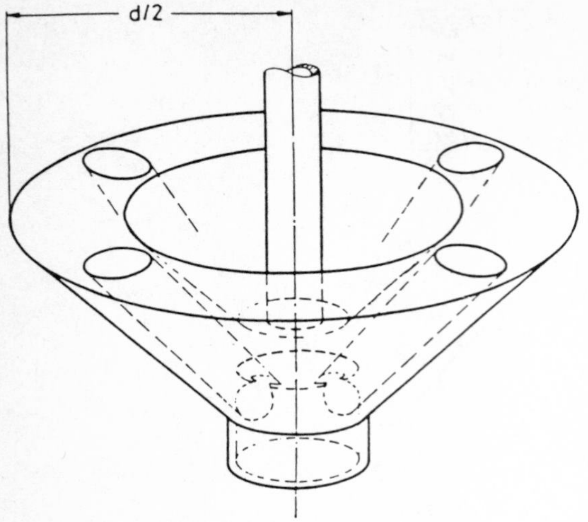

Fig. A 7.1: Drawing of the mechanical foam breaker

centrifugal force to gently compress foam. Our task will be to perform model measurements in such a way as to obtain data for design and scale-up of these devices.

However, before starting model tests, *preliminary* tests will have to be carried out to give us an idea of the relevance list of the problem. These will show that, in the given material system, a minimum rotational speed n of the foam breaker with the diameter d is necessary to control the foam rise at each gas throughput q. (The task is to prevent a discharge of the off-gas from the vessel as foam; the head space above the liquid in contrast can be filled with a random quantity of foam.)

The physical properties of the foam constitute a problem: They are either unknown or they cannot be measured. (The elasticity and viscosity of the liquid lamellae are dynamic quantities; furthermore, the concentration of surfactants in the lamella varies during bubble formation according to Gibbs´ law.) Therefore, when compiling the relevance list, we will have to use S_i to globally designate the relevant physical properties of the foam:

$\{ n, d; q; S_i \}$

This relevance list consists of *two* process parameters (n, q), one of which (the minimum rotational speed n) is the *target* quantity; a characteristic *linear dimension* d and an unknown number of partly unknown *physical properties* S_i. We will use these to form a series of material numbers S_j^* and two process numbers. Of these two the target number can be formulated instantly: $Q = n\, d^3/q$, while the second process number will be made up of the elements $(q, d, S_k)^*$.

The Π set will therefore be:

$$\{\, Q \equiv nd^3/q\,,\ (q, d, S_k)^*,\ S_j^*\,\}$$

In order to work out the structure of the process number and, consequently, the process characteristic of the foam breaker, it is necessary to perform tests in *differently sized* models, initially using the *same material* system.

Three geometrically similar models (d = 200; 300; 400 mm) were available. It was possible to operate the foam breaker with a variable rotational speed. The gas flow was distributed in the liquid with a turbine stirrer, the volume-related stirrer power ($P/V = 500\ W/m^3$) of which was adapted to the gas throughput in each case. The same surfactant was used. The air throughput q was increased in steps and the *lowest* rotational speed of the foam breaker, with which it was just possible to control the foam, was set for each gas throughput. In this way the pairs of (n, q) were established in each model (at d = const). The dimensionless number Q was then calculated from these values and plotted over the corresponding gas throughputs q in the double-logarithmic plot, see Fig. A 7.2.

Three parallel straight lines were obtained in this way for three differently sized models. Correlation of these results was possible with the parameter d leading to the following relationship (Fig. A 7.3):

$$Q = f_1(q/d^{2,5}) \qquad \text{or} \qquad Q = f_2(q^2/d^5)$$

These findings were confirmed by measurements with other surfactants [B 13]; the structure of this process variable (q^2/d^5) can therefore be considered as verified. It has the dimension of an acceleration $[L\, T^{-2}]$ and can

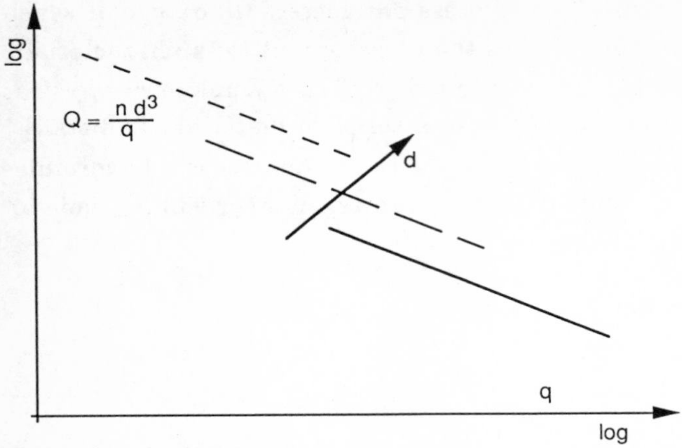

Fig.A 7.2: Presentation of the test results obtained with three differently sized models (d = 200; 300; 400 mm) in the form Q(q)

be made dimensionless with relevant physical properties S_k in a suitable combination.

Of course it is unsatisfactory to have a process characteristic such as that shown in Fig. A7.3 in which the abscissa exhibits a dimensional process variable. However, this disadvantage can be easily rectified[22] even

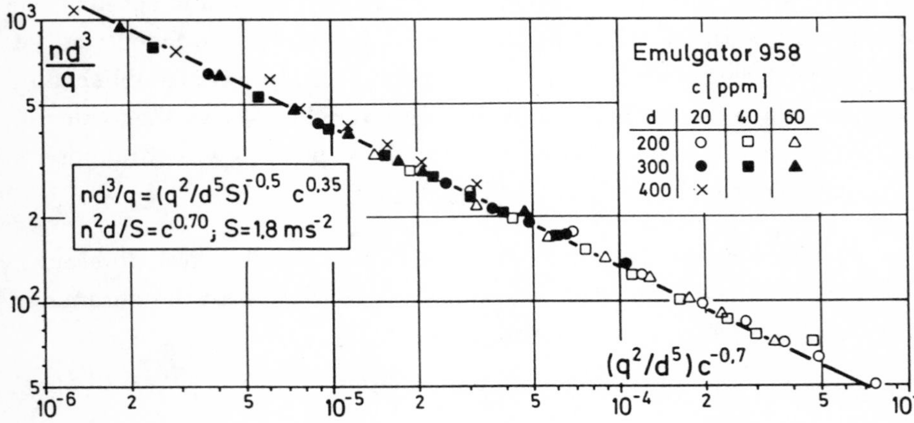

Fig. A 7.3: Correlation of the test results obtained for three models with the process variable q^2/d^5. Taken from [B13]

---

22) I am obliged to *Dr. J. Pawlowski* for this idea

without knowledge of S_k if the relationship in Fig. A 7.3 is converted as follows:

$$n\,d^3/q = \text{const}\,(q^2/d^5)^{\alpha} \quad \rightarrow \quad n\,d^3/q = 1\,(q^2/d^5\,S)^{\alpha}$$

whereby the numerical value of the unknown variable S [L T^{-2}] results from the ordinate intercept (const) and the exponent (α):

$$S\,[\text{m s}^{-2}] = \text{const}^{-1/\alpha}$$

The variable S can be taken as an expression of foam breakability and the described testing procedure and evaluation, which were previously employed to establish the process characteristic of the foam breaker, can be used as a method for determining foam stability. The results of tests with 5 different surfactants at different concentrations are presented in Fig.A 7.4. This shows that foam stability is dependent on both the type of surfactant and its concentration.

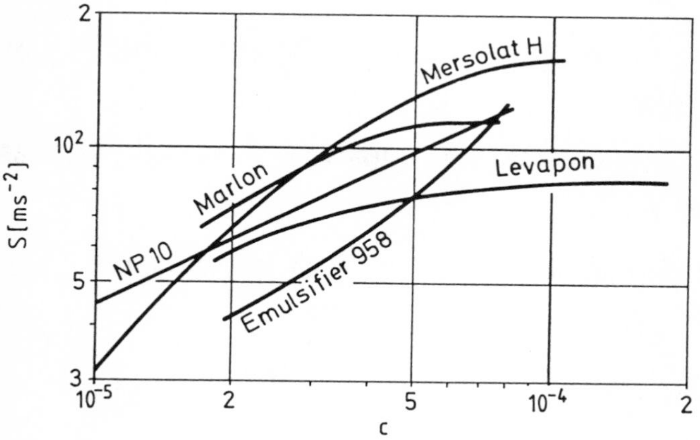

Fig.A 7.4: S [ms^{-2}] as a function of the type and concentration [ppm] of the surfactant. Taken from [B13]

Example A 8:

Description of the temporal course of spin drying in centrifugal filters
(taken from [B 14])

This example deals with constant-speed batch centrifugal filters operating on a horizontal axis. The operating cycle is loading, wet spin, cake wash, dry spin, unloading (peel out). Dry spin requires most time. It consists of the rapid draining of the mother liquor from the capillary spaces and the slow draining of the surface liquor. The dry spin which governs the flowrate has been completed when the equilibrium residual moisture w_∞ is attained in the filter cake.

Before considering the dry spin process in a dimensional analysis, some terms have to be explained and defined.

1. *Centrifugal acceleration* b [LT^{-2}] is expressed by the multiple (z) of gravitational acceleration g: $b = z\,g$

2. The *specific filter cake resistance* α [L^{-2}] is defined with the equation describing the pressure loss Δp of the liquid in the porous filter cake at laminar flow:
$\Delta p = \alpha\, v\, \eta\, h$
($v \equiv q/A$ – liquid flowrate q related to the filter surface A; h – cake height; η – dynamic viscosity)

3. The *porosity* ε [–] of the filter cake is defined as the ratio of pore volume to total volume.

4. The *residual moisture* w [–] of the filter cake reflects the ratio of liquid mass to solids mass.

5. The *degree of saturation* S [–] is defined as the ratio of the pore volume filled with liquid to the total pore volume:

$$S = w\,\frac{\rho_S}{\rho_W}\,\frac{(1-\varepsilon)}{\varepsilon} = \frac{w}{w_{max}}$$

whereby ρ_S and ρ_W are the densities of solid matter and water respectively and w_{max} is the cake moisture at saturation.

Equilibrium saturation of the cake $S_\infty \equiv w_\infty / w_{max}$ will initially depend on the physical properties of the filter cake. They are characterized by α, ε, Θ and K. Θ represents the contact angle (degree of wetting!) and K any further parameters of the grain such as roughness etc. Furthermore, the physical properties of the wash liquid (density ρ and surface tension σ)

and, finally, centrifugal acceleration b as process parameter will be of importance:

$\{ S_\infty, \alpha, \varepsilon, \Theta, K, \rho, \sigma; b \}$

Four of these eight process-relevant variables are dimensionless, the other four form one further dimensionless number:

	ρ	α	b	σ
M	1	0	0	1
L	−3	−2	1	0
T	0	0	−2	−2

Z_1	1	0	0	1	$Z_1 = M$
Z_2	0	−2	1	3	$Z_2 = L + 3M$
Z_3	0	0	1	1	$Z_3 = -1/2\,T$

Z'_1	1	0	0	1	$Z'_1 = Z_1$
Z'_2	0	−2	0	2	$Z'_2 = Z_2 - Z_3$
Z'_3	0	0	1	1	$Z'_3 = Z_3$

Z''_1	1	0	0	1	$Z''_1 = Z_1$
Z''_2	0	1	0	−1	$Z''_2 = -1/2\,Z'_2$
Z''_3	0	0	1	1	$Z''_3 = Z_3$

$$\boxed{\Pi_1 \equiv \frac{\sigma\,\alpha}{\rho\,b}}$$

It follows that:

$$S_\infty \equiv w_\infty / w_{max} = f(\Pi_1, \varepsilon, \Theta, K)$$

Tests [B 14] have shown that this relationship is described by the analytical expression

$$S_\infty = \Pi_1^{0.2}\, f(\varepsilon, \Theta, K) = \left(\frac{\sigma\,\alpha}{\rho\,b}\right)^{0.2} f(\varepsilon, \Theta, K)$$ and hence by

$$S_\infty = \text{const}\,(1/z)^{0.2}$$ whereby $z = b/g$.

The numerical values of the constants and the exponent are naturally dependent on the material system under examination.

In order to track the time course of the dewatering process up to an average degree of saturation $S_{tm} \equiv w_{tm}/w_{max}$, the parameters time t, viscosity η of the wash liquid and the geometric parameters of the cake (cake height h and cake residual height h_o remaining after peel out) must be added to the above relevance list. Since we are dealing with a creeping flow movement in the centrifugal field, ρ only takes effect in combination with b: ρb; compare the form of Π_1! Apart from the obvious geometric numbers h/h_o and αh^2 and the dimensionless parameters S_{tm}, ε, Θ, K, two further numbers will be involved:

	ρb	a	t	s	h
M	1	0	0	1	1
L	−2	−2	0	0	−1
T	−2	0	1	−2	−1

Z_1	1	0	0	1	1	$Z_1 = M$
Z_2	0	−2	0	2	1	$Z_2 = L + 2M$
Z_3	0	0	1	0	1	$Z_3 = T + 2M$

Z'_1	1	0	0	1	1	$Z'_1 = Z_1$
Z'_2	0	1	0	−1	−1/2	$Z'_2 = -1/2\, Z_2$
Z'_3	0	0	1	0	1	$Z'_3 = Z_3$

$$\Pi_1 \equiv \frac{\sigma\,\alpha}{\rho b} \qquad \Pi_2 \equiv \frac{\eta\,\alpha^{1/2}}{\rho b\, t}$$

Π_1 is the same number as that formed before. The complete Π set is now:
$\{ S_{tm}, h/h_o, \alpha h^2, \varepsilon, \Theta, K, \Pi_1, \Pi_2 \}$

The tests [B 14] were performed with small acryl glass spheres of δ = 20 - 50 µm in diameter. The material numbers ε, Θ, K remained unchanged. However, h/h_o, αh^2, Π_1 and Π_2 were varied by changing b, t and h. It was found that the test results can be correlated in the Π space $\{S_{tm}, \alpha h^2, \Pi_2\}$, i.e., neither Π_1 nor h/h_o is significant. Fig. A 8.1 shows the result. The reciprocal value of $\Pi_2 (\alpha h^2)^{0.5} = \{\rho b\, t /(\eta\, \alpha\, h)\}^{-1}$ is plotted on the abscissa. The process relationship is:

$$S_{tm} = 0.26\, \{\rho b\, t /(\eta\, \alpha\, h)\}^{-2/3}$$

The non-relevance of h_o is not surprising if the solid particles are neither damaged nor compressed in the peel out process and if the capillary rise height is << h_o. The above Π space should also apply for filter cakes which can be compressed to a greater extent [B 14]. The fact that Π_1 is not significant at total wetting can only document the irrelevance of σ during the dry spin process.

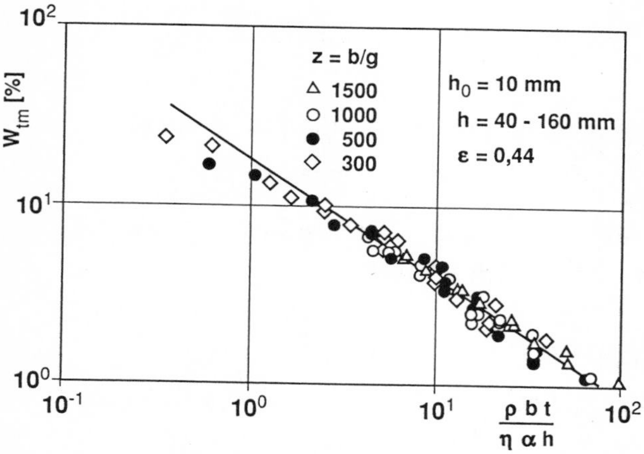

Fig. A 8.1: Temporal course of the average residual moisture w_{tm} as a function of the dimensionless number $\{\rho b\, t/(\eta\, \alpha\, h)\}$ for given geometric and material conditions at centrifugal accelerations of b = 300 - 1500 g. Taken from [B 14]

Example A 9:

Description of particle separation by means of inertial forces

(taken from [B 15])

Let us consider the separation of aerosols (droplet size $\delta = 0.2 - 20\ \mu m$) from a gas flow in a dust separator (e.g., wire filter, cyclone, etc.). The result – the *fractional degree* η_F – is characterized by

$$\boxed{\eta_F = (\varphi_{in} - \varphi_{out})/\varphi_{in}}\ .$$

This target quantity is dependent on the following quantities:

Geometric parameters: Particle diameter δ and a characteristic linear dimension D of the separator.

Physical properties: Particle density ρ_P,
Density ρ and viscosity η of the gas

Process parameters: Gas velocity v *or better*: the pressure drop $\Delta p \sim v^{\alpha}$, ($\alpha = 1$ to 2) because this variable is characteristic of the separating device.

To summarize:

$$\boxed{(\eta_F; \delta, D; \rho_P, \rho, \eta, \Delta p)}\ .$$

A closer examination of the problem and the evaluation of the preliminary tests can streamline this 7-parametric relevance list:

1. *Stokes* law applies for sufficiently small particles. According to this, the *frictional force* is $F_{fr} = 6\,\pi\, r\, \eta\, v = 3\,\pi\, d\, \eta\, v$

At the same time, the *inertial force* is $F_{mass} = m\, b = (\pi/6)\, \rho_P\, \delta^3\, b$

In the steady-state $F_{fr} = F_{mass}$. For the *settling velocity*, it follows that $v = \rho_P\, \delta^2\, b/(18\, \eta)$. From this we register that $v \sim \rho_P\, \delta^2$

2. The gas velocity v and thus the pressure drop Δp which is necessary for separation of a specific limiting droplet size (e.g., δ_{50}) decreases with δ and is strongly dependent on the type of separator device:

$$\Delta p \sim \delta^{-3} \quad \rightarrow \quad \Delta p\, \delta^3 = \text{const (geometry)}.$$

Since the particle size is bound on the one hand as $\rho_P\, \delta^2$ and on the other hand as $\Delta p\, \delta^3$, we will combine these two requirements to obtain:

$$\rho_P^3\, \delta^6\, \Delta p^2$$

As a result of closer examination (1) and the preliminary tests (2), the above 7-parametric relevance list can be reduced to the following 5-parametric list:

$$\boxed{(\eta_F; D; \rho, \eta; \rho_P^3\, \delta^6\, \Delta p^2)}$$

Since η_F is the target number, we only have to form one single process number:

	ρ	D	η	$\rho_P^3\,\delta^6\,\Delta p^2$
M	1	0	1	5
L	−3	1	−1	−5
T	0	0	−1	−4

Z_1	1	0	1	5	$Z_1 = M$
Z_2	0	1	2	10	$Z_2 = L + 3M$
Z_3	0	0	1	4	$Z_3 = -T$

Z'_1	1	0	0	1	$Z'_1 = Z_1 - Z_3$
Z'_2	0	1	0	2	$Z'_2 = Z_2 - 2\,Z_3$
Z'_3	0	0	1	4	$Z'_3 = Z_3$

$$\frac{\rho_P^3\,\delta^6\,\Delta p^2}{\rho\,D^2\,\eta^4}$$

We will first draw the square root of this dimensionless group and then relate it to the well-known Euler, Reynolds and Stokes numbers:
$Eu \equiv \Delta p/(\rho\,v^2)$; $Re \equiv v\,D\,\rho/\eta$; $Stk \equiv \rho_P\,\delta^2\,v/(D\,\eta)$.

It follows that

$$A \equiv \frac{\rho_P^{3/2}\,\delta^3\,\Delta p}{\rho^{1/2}\,D\,\eta^2} \equiv Eu\,Re^{1/2}\,Stk^{3/2}$$

Bürkholz [B 15] calls this combination of dimensionless numbers the "deposition number" $\Psi_A^{3/2}$. Its structure is identical to A.

The significance of this "deposition number" is impressively demonstrated by the comparison in Fig. A 9:

Furthermore, the dimensionless number $A \equiv \Psi_A^{3/2}$ indicates the existence of the relationship $\Delta p \sim D$. For each desired fractional degree of separation, the necessary pressure drop Δp is proportional to the characteristic linear dimension D. Different separator types can therefore be compared if D is selected sensibly, see [B 15].

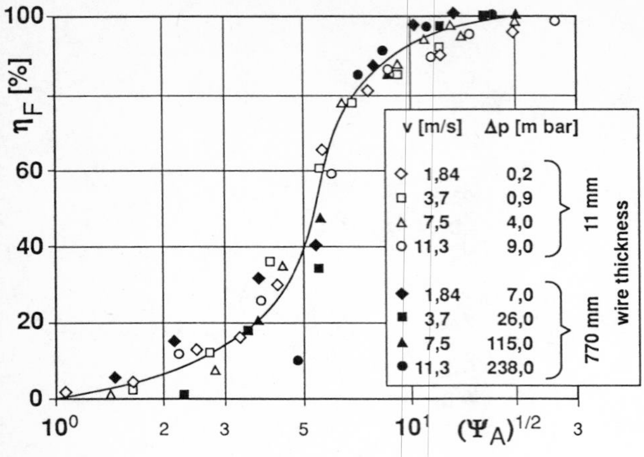

Fig. A 9: Fractional degree of separation n_F of two differently thick wire filters (11 and 770 mm) having the same wire thickness. Correlation of the measurements by means of the deposition number $A \equiv \Psi_A{}^{3/2}$. Taken from [B 15].

Example A 10:

Conveying characteristics of single-screw machines for Newtonian and non-Newtonian liquids. Optimum conditions (P/q = min) and scale-up

Screw machines are frequently used for the production (mass polymerization) and processing (mixing, extrusion) of plastics. A distinction is made between single-screw and multiple-screw machines (e.g., self-wiping twin-screws[23)]) and between conveying, mixing and kneading screw machines.

The following characteristics [A 9] give the conveying properties of a single-screw machine of given screw geometry in the creeping motion range (Re < 100) of Newtonian liquids:

[23)] R. Erdmenger, Chem.-Ing.-Tech. **36** (1964) 3, 175/185: "Mehrwellen-Schnecken in der Verfahrenstechnik" (Multiple screw machines in chemical process engineering)

Pressure characteristic: $\quad \mathrm{Eu\,Re\,d/L} \equiv \dfrac{\Delta p\, d}{\eta\, n\, L} = f_1(Q)$

Axial force characteristic: $\quad \mathrm{Ne_F\,Re\,d/L} \equiv \dfrac{F}{\eta\, n\, d\, L} = f_2(Q)$

Power characteristic: $\quad \mathrm{Ne_F\,Re\,d/L} \equiv \dfrac{P}{\eta\, n^2\, d^2\, L} = f_3(Q)$

Q represents the flow rate number $Q \equiv q/(nd^3)$. These three characteristics are illustrated in Fig.A10.1 for a screw of given geometry. They are linear dependences which are described by analytical expressions in the form

$\dfrac{1}{y_1}Y + \dfrac{1}{q_1}Q = 1$ $\quad$ (y_1 and q_1 are the respective axis intercepts).

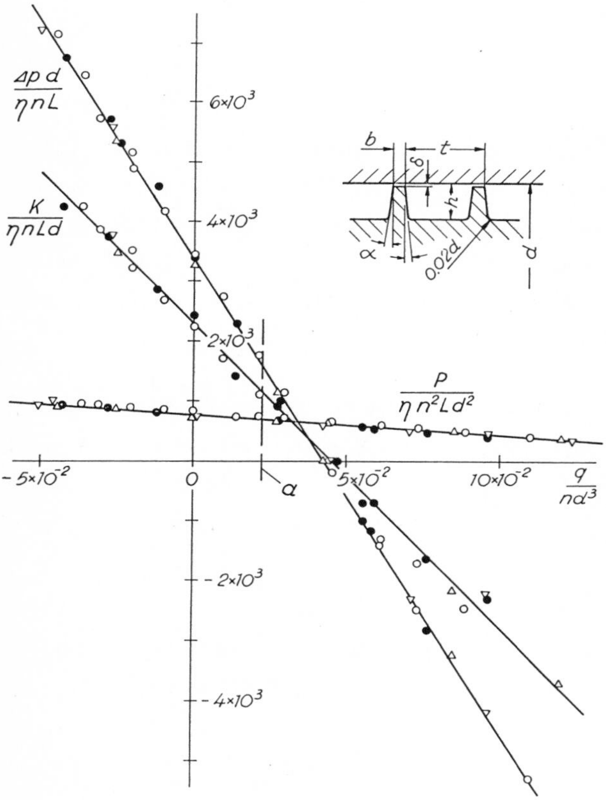

Fig. A 10.1: Conveying characteristics of a single-screw machine of given screw geometry. Taken from [A 9].

Of the three ranges of the pressure characteristic, only the so-called active conveying range of the screw machine $0 < Q < q_1$ can be implemented by suitable throttling and/or a change in the rotational speed alone. In order to implement the other two ranges it is necessary to couple the screw machine with an additional conveying device (e.g.,a positive displacement gear-type rotary pump). If the pump transports the liquid in the same direction as the screw, the range $Q > q_1$ results. At $Q < 0$ the pump pushes the liquid against the conveying sense of the screw. The latter is then only a mixing device.

In the case of a non-Newtonian liquid, such as that for which the viscosity curve is given in Fig. 6, the pressure characteristic is depicted in the Π set:

$$\{ \Delta p\, d/(H\, n\, L),\ Q,\ n\Theta,\ \Pi_{rheol} \}$$

Fig. A 10.2 shows this relationship for the aforementioned liquid; it was established in two differently sized (d = 60 and 90 mm) single-screw machines with the same profile geometry using two rotational speeds (n = 1.65 and 25 min^{-1}). The higher the rotational speed, the higher is the shear stress; the straight line (a) which is also valid for Newtonian liquids adjusts as the limit case ($\eta = \eta_\infty$).

Operating conditions at which a minimum of power is dissipated in the liquid (= lowest thermal load of the liquid throughput) are frequently of interest. The dissipated power H is obtained from the difference between the power P of the motor drive and that of the pump $q\Delta p$:

$$H = P - q\Delta p.$$

A dimensionless formulation of this relationship is possible using the conveying characteristics of the screw machine in question, see Fig.
A 10.1. In the active conveying range

$$0 < Q < q_1 \qquad \text{resp.} \qquad 0 < \Lambda \le 1 \qquad \text{with} \quad \Lambda \equiv Q/q_1$$

the dissipation characteristic passes through a minimum, in which the lowest power dissipation H occurs for the given values of q and Δp, corresponding to H/q = min.

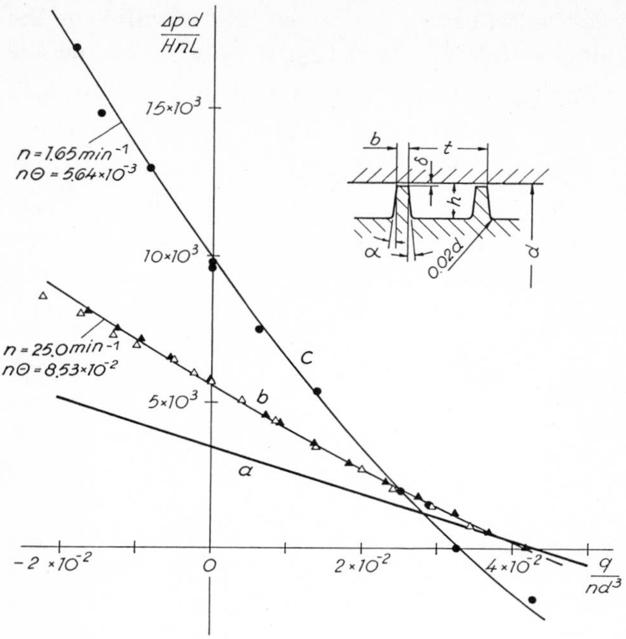

Fig. A 10.2: Pressure characteristic of a screw machine of given screw geometry for the non-Newtonian liquid according to Fig. 6. Taken from [A 9].

In Sect. 4.3, p. 55 it was pointed out that, even using the same non-Newtonian liquid, complete similarity of the model and the full-scale counterpart can only be attained if there is a creeping, steady-state and isothermal flow condition. Scale-up is then carried out as follows:

The non-Newtonian liquid and the parameters Δq_0 and p_0 of the industrial facility are given. We are searching for the variable d and the rotational speed n of the prototype, whereby P = min is required. Corresponding to the above Π set for non-Newtonian liquids, it follows that:

$$L/d;\ n;\ \Delta p = \text{idem} \qquad \rightarrow \qquad q/d^3;\ P/q = \text{idem.}$$

In the model screw machine, the dependences q(n) and P(q) are established at $\Delta p = \Delta p_o$ and depicted as $P/q = f_1(n)$ and $q/d^3 = f_2(n)$. This situation also applies to the prototype. The possibly existing minimum of

P/q gives the optimum speed of rotation n_{opt}, which is also valid for the prototype. The corresponding values d_{opt} and P_{opt} for the prototype are obtained from the values $(q/d^3)_{opt}$ and $(P/q)_{opt}$ by setting $q = q_o$. We have therefore solved the task.

B Examples from the Field of Thermal Unit Operations – Heat and Mass Transfer

Introductory Remarks

Besides fluid mechanics thermal processes also include *mass transfer* processes (e.g. absorption or desorption of the gas in the liquid, extraction between two liquid phases, dissolution of solids in liquids) and/or *heat transfer* processes (energy uptake, cooling, heating). In the case of thermal separation processes such as distillation, rectification, drying etc., mass transfer between the phases is subject to *thermodynamic* laws which are not scale-dependent. Therefore one should not be surprised if there are no scale-up rules for the pure rectification process unless the hydrodynamics of the mass transfer in plate and packed columns are under consideration. If a separation operation, e.g., electrophoresis (Example B6), involves *simultaneous* mass and heat transfer, both of which are scale-dependent, scale-up is particularly difficult because the two processes obey different laws.

Heat transfer processes are described by physical properties and process-related parameters, the dimensions of which include not only the primary quantities of **M**ass, **L**ength and **T**ime but also temperature Θ as the fourth one. Examples B1 and B3 show that, in the dimensional analysis of such problems, it is advantageous to expand the dimensional system to include heat **H** [kcal] as the fifth primary quantity. Joule's mechanical equivalent of heat J must then be introduced as the corresponding dimensional constant in the relevance list. Although this does not change the Π space, a dimensionless number is formed which contains J and, as such, frequently proves to be irrelevant. As a result, the Π set is finally reduced by one dimensionless number, c.f. Sect. 2.5 on p. 38.

Example B1:

Steady-state heat transfer in the mixing vessel at cooling and the optimum conditions for maximum removal of the heat of reaction

The heat flux Q (quantity of heat per unit of time) through a heat exchanger surface A is described by the overall heat transfer equation

$$\boxed{Q = k\, A\, \Delta T}$$

where ΔT is a characteristic temperature difference between the surface (wall) and the liquid. The proportionality factor is the overall heat transfer coefficient k

$$\boxed{\frac{1}{k} = \frac{1}{\alpha_a} + \frac{\delta}{\lambda} + \frac{1}{\alpha_i}}$$

i.e. the heat transfer resistances are additive terms.

(α_a and α_i are the heat transfer coefficients at the outside and inside of the heat exchanger surface; λ and δ are the heat conductivity and the wall thickness)

In operation of a mixing vessel for the given material system, one is interested in mixing conditions which change the heat transfer coefficient α_i – referred to as α in the following. This target variable is dependent on the following parameters:

Geometric parameters:	Vessel and stirrer diameter: D, d;
Physical properties:	Density ρ ; viscosity η ; temperature coefficient of viscosity γ (see p. 53), heat capacity c and heat conductivity λ of the liquid
Process parameters:	Rotational speed n of the stirrer, temperature difference ΔT between wall and liquid.

The complete 10 parametric relevance list is as follows:

$\{\alpha;\ D,\ d;\ \rho,\ \eta,\ \gamma,\ c,\ \lambda;\ n,\ \Delta T\}$

In the following we will subject it to a dimensional analysis in conjunction with both dimensional systems coming into question here, namely {M, L, T, Θ} and {M, L, T, Θ, H} (see Table 2 on p. 17). In the case of the second dimensional system, Joule's equivalent of heat *J* must be included in the relevance list.

Dimensional system {M, L, T, Θ} :

	ρ	d	n	ΔT	α	c	λ	η	D	γ
M	1	0	0	0	1	0	1	1	0	0
L	−3	1	0	0	0	2	1	−1	1	0
T	0	0	−1	0	−3	−2	−3	−1	0	0
Θ	0	0	0	1	−1	−1	−1	0	0	−1

Z_1	1	0	0	0	1	0	1	1	0	0	$Z_1 = M$
Z_2	0	1	0	0	0	2	1	−1	1	0	$Z_2 = L + 3M$
Z_3	0	0	1	0	−3	−2	−3	−1	0	0	$Z_3 = -T$
Z_4	0	0	0	1	−1	−1	−1	0	0	−1	$Z_4 = \Theta$

The following 6 dimensionless numbers are obtained:

$$\Pi_1 \equiv \frac{\alpha\, \Delta T}{\rho\, d^3\, n^3} \qquad \Pi_4 \equiv \frac{\eta}{\rho\, d^2\, n} \equiv Re^{-1}$$

$$\Pi_2 \equiv \frac{c\, \Delta T}{d^2\, n^2} \qquad \Pi_5 \equiv D/d$$

$$\Pi_3 \equiv \frac{\lambda\, \Delta T}{\rho\, d^4\, n^3} \qquad \Pi_6 \equiv \gamma\, \Delta T$$

However, the appearance of the numbers Π_1, Π_2 and Π_3 is unknown to us. First of all, we will recombine the α-containing target number Π_1 with Π_3:

$$\Pi_1\, \Pi_3^{-1} \equiv \alpha\, d/\lambda \;\rightarrow\; \alpha\, D/\lambda \equiv Nu \qquad \text{(Nu – Nusselt number).}$$

Then, using Π_3 and Π_4, we will convert Π_2 into a pure material number:

$$\Pi_2\, \Pi_3^{-1}\, \Pi_4^{-1} \equiv c\, \eta/\lambda \equiv Pr \qquad \text{(Pr – Prandtl number).}$$

The only number we do not know is Π_3. Furthermore, we are not aware of its importance.

Therefore dimensional analysis with the dimensional system {M, L, T, Θ} gives the following 6-parametric Π set for the heat transfer characteristic of a mixing vessel:

$$\boxed{\{\, \mathrm{Nu}, \mathrm{Pr}, \Pi_3, \gamma\Delta T, \mathrm{Re}, D/d \,\}}$$

whereby the importance of Π_3 is still not apparent.

<u>Dimensional system {M, L, T, Θ, H}</u>

	ρ	d	n	ΔT	λ	α	c	J	η	D	γ
M	1	0	0	0	0	0	−1	1	1	0	0
L	−3	1	0	0	−1	−2	0	2	−1	1	0
T	0	0	−1	0	−1	−1	0	−2	−1	0	0
Θ	0	0	0	1	−1	−1	−1	0	0	0	−1
H	0	0	0	0	1	1	1	−1	1	1	0

Z_1	1	0	0	0	0	0	−1	1	1	0	0	$Z_1 = M$
Z_2	0	1	0	0	−1	−2	−3	5	2	1	0	$Z_2 = L + 3M$
Z_3	0	0	1	0	1	1	0	2	1	0	0	$Z_3 = -T$
Z_4	0	0	0	1	−1	−1	−1	0	0	0	−1	$Z_4 = Q$
Z_5	0	0	0	0	1	1	1	−1	0	0	0	$Z_5 = H$

Z'_1	1	0	0	0	0	0	−1	1	1	0	0	$Z'_1 = Z_1$
Z'_2	0	1	0	0	0	−1	−2	4	2	1	0	$Z'_2 = Z_2 + Z_5$
Z'_3	0	0	1	0	0	0	−1	3	1	0	0	$Z'_3 = Z_3 - Z_5$
Z'_4	0	0	0	1	0	0	0	−1	0	0	−1	$Z'_4 = Z_4 + Z_5$
Z'_5	0	0	0	0	1	1	1	−1	0	0	0	$Z'_5 = Z_5$

The following dimensionless numbers are obtained:

$$\Pi_1 \equiv \frac{\alpha\, d}{\lambda};\ \Pi_2 \equiv \frac{c\,\rho\, d^2\, n}{\lambda};\ \Pi_3 \equiv \frac{J\,\Delta T\,\lambda}{\rho\, d^4\, n^3};\ \Pi_4 \equiv \frac{\eta}{\rho\, d^2\, n};\ \Pi_5 \equiv \frac{D}{d};\ \Pi_6 \equiv \gamma\Delta T.$$

$\Pi_1 \rightarrow \mathrm{Nu} \equiv \alpha\, D/\lambda$ (Nu – Nusselt number)

$\Pi_2\Pi_4 \equiv c\,\eta/\lambda \equiv \mathrm{Pr}$ (Pr – Prandtl number)

$$\Pi_3^{-1}\,\Pi_4 \equiv \frac{\eta\, d^2\, n^2}{J\,\Delta T\,\lambda} \equiv \mathrm{Br} \qquad \text{(Br – Brinkman number)}$$

$$\Pi_4 \equiv \mathrm{Re}^{-1} \qquad \text{(Re – Reynolds number)}$$

In both dimensional analyses, the dimensionless number Π_3 was obtained. (After combination with the Re number it can be interpreted as the *Brinkman* number.) Because it contains J we realize that it takes into account the heat production of the mixing device! However, as long as heat production is negligible in comparison to heat removal, Π_3 resp. Br is irrelevant and can be eliminated. The complete Π set is therefore:

$$\boxed{\{\mathrm{Nu}, \mathrm{Pr}, \mathrm{Re}, D/d, \gamma\Delta T\}}\ .$$

Although both dimensional analyses lead to the same result with regard to the numbers formed, the second permits better interpretation because of the occurrence of J in one of the numbers and therefore allows a decision to be made about their relevance, c.f. Chapt. 2.5 on p. 38.

Fig. B1.1 shows the results of measurements [B 16] of steady-state heat transfer at cooling in a laboratory mixing vessel with anchor stirrer (D = 400 mm; D/d = 1.02). The liquids used were selected in such a way that not only a wide Re range was covered; the temperature coefficient of viscosity γ was also varied substantially. As is apparent from the (lower) depiction of the results in the turbulent regime $\mathrm{Re} > 10^3$, the influence of the rheological number $\gamma\Delta T$ is very slight. The process equation is valid for this range:

$$\boxed{\mathrm{Nu}\,\mathrm{Pr}^{-1/3} \sim \mathrm{Re}^{2/3}}\ .$$

If the presentation of the measurement results in the form Nu(Re) (*upper* part of the figure) is compared to that in the plot $\mathrm{Nu}\,\mathrm{Pr}^{-1/3}(\mathrm{Re})$ (*lower* part of the figure), it is apparent that the traditional consideration of the material number Pr only has a correlating effect in the turbulent flow regime while, in the laminar flow regime ($\mathrm{Re} < 10^2$), it even exhibits a deterioration as compared to Nu(Re). Since a further dimensionless number is not at hand, this shows that the influence of Pr in this range must be considered in a different way. The following consideration is helpful:

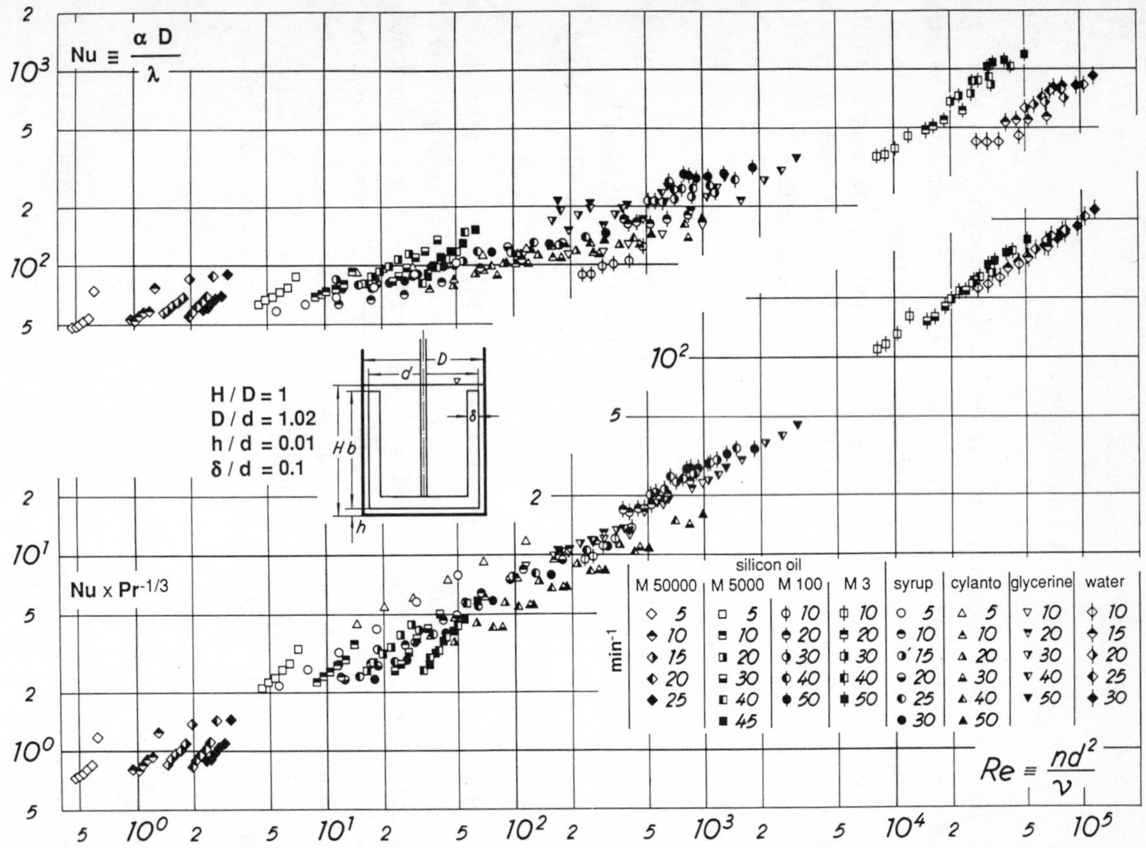

Fig. B1.1: Steady-state heat transfer at cooling in a mixing vessel with anchor stirrer (geom. parameters: see sketch). Observe the influence of the Pr number in the entire Re range. [B 16]

In the *turbulent* regime the correlation applies:

$Nu \sim Re^{2/3} Pr^{1/3} = (Re\, Pr^{1/2})^{2/3}$.

Assuming that the ratio of the exponents of Re and Pr has a *constant* value of 1 : 0.5 in the *entire Re range*, the following correlation could be expected in the range of *creeping flow*:

$Nu \sim Re^0 Pr^0 = (Re\, Pr^{1/2})^0$.

In fact, a satisfactory correlation of all experimental material can be attained in this framework [B 16, Fig. 5]. The process equation with the measurement results of Fig. B1.1 is then:

$$\boxed{Nu = 0.23\,(Re\, Pr^{1/2} + 4000)^{2/3}}$$

However, an increase in the rotational speed not only improves the heat transfer coefficient α - which increases the heat flux Q through the vessel wall - , it also enhances the stirrer power P which results in heat of agitation, see Fig.B1.2. This means that designing a given stirrer for a cooling process (e.g., for the removal of heat of reaction per unit of time, R) is an *optimization task*, namely the determination of the optimum rotational speed n_{opt} which guarantees a maximum of $R = Q - P$.

This task can be simplified with the aid of dimensional analysis as shown in the following, using the above result which is valid for the anchor stirrer in the laminar flow regime (Re < 100). In this flow range it can be assumed that $k \approx \alpha_i$, i.e. the heat transfer coefficient α_i on the inside of the vessel wall is the determinative parameter.

The question as to how much heat of reaction R can be removed in a certain volume V:

$W \equiv R/V = (Q - P)/V = (\alpha\, A\, \Delta T/V) - P/V$

is formulated in dimensionless form as follows:

$$\boxed{\frac{W D^2}{\lambda\, \Delta T} = Nu - (D/d)\frac{\eta^3}{D^2 \rho^2 \lambda\, \Delta T}\, Re^3\, Ne}$$

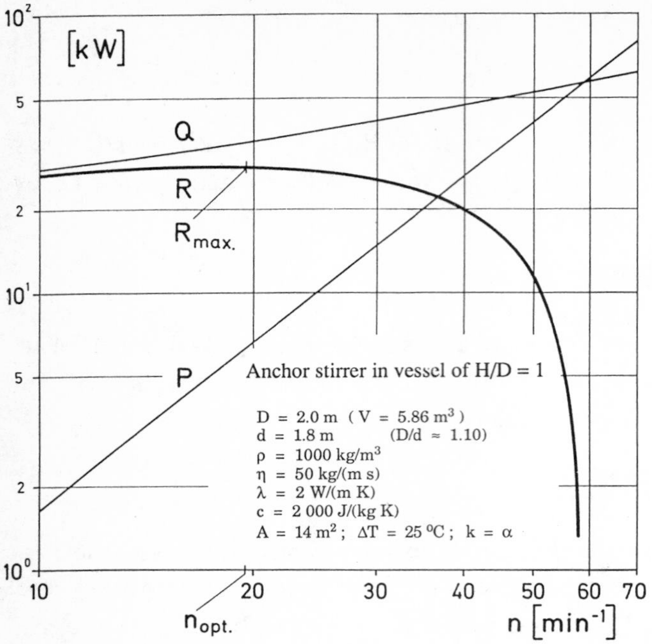

Fig. B1.2: Dependence of the removable heat Q and the heat of agitation P on the rotational speed n under the given operating conditions

We will use the following abreviations:

$$\Pi_1 \equiv \frac{\eta^3}{D^2 \rho^2 \lambda \Delta T} \qquad \text{and} \qquad \Pi_2 \equiv \frac{W D^2}{\lambda \Delta T}$$

Since Nu = Nu(Re,Pr) and Ne = Ne(Re) are known functions, $\Pi_2 = f$ (Re, Pr, Π_1). Optimum operation is characterized by the maximum of Π_2 in relation to Re and can therefore be determined by differentiating f with respect to Re and setting the derivative to zero. For ease of handling, the result is presented graphically in Fig. B1.3. It is valid for the anchor stirrer for two D/d ratios and for Re < 100. Details relating to the derivation of this depiction are given in [B 17].

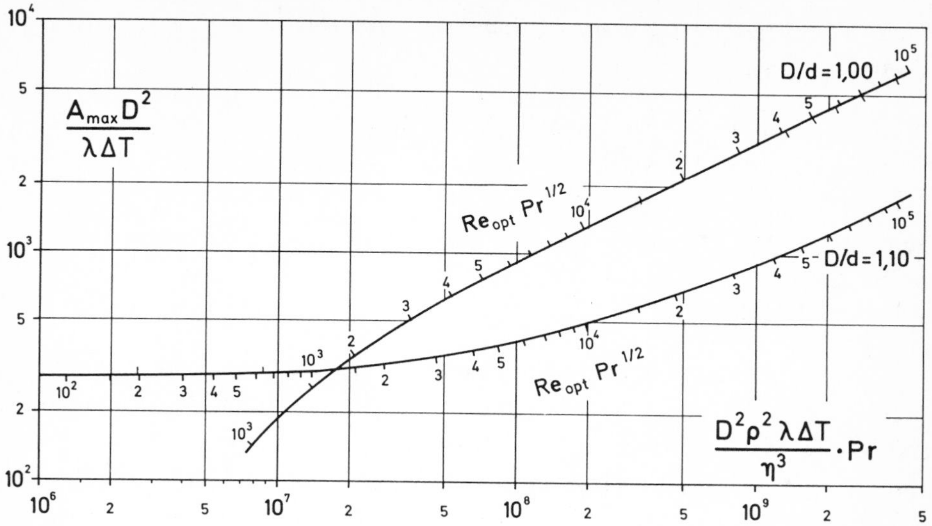

Fig. B1.3: Working sheet to determine the mixing conditions for maximum removal of the heat of reaction in the range Re < 100. Taken from [B 17].

Example B2:

Steady-state heat transfer in bubble columns

After dealing extensively with steady-state heat transfer in mixing vessels in Example B1, it is easy to compile the relevance list for this process in bubble columns, especially since hydrodynamics in bubble columns were considered by means of a dimensional analysis in Example A5.

Target quantity:	Heat transfer coefficient α
Geom. parameter:	Column diameter D
Physical properties:	Density ρ and viscosity η of the liquid
	Heat capacity c and conductivity λ,
Process parameters:	Gas throughput q
	Gravity difference $g\Delta\rho$ between gas and liquid.

$\{ \alpha; D; \rho, \eta, c, \lambda; q, g\Delta\rho \}$

This 8-parametric set can be converted into the following four dimensionless numbers:

Nusselt number	$Nu \equiv \alpha D/\lambda$
Prandtl number	$Pr \equiv c\,\eta/\lambda$
Reynolds number	$Re \equiv q\,\rho/(D\,\eta)$
Froude number	$Fr^* \equiv q^2\,\rho/(D^5\,g\Delta\rho)$

$\{ Nu, Pr, Re, Fr^* \}$

W. Kast [B18] discovered that the intensity variable superficial velocity v ~ q/D^2 is decisive for heat transfer in bubble columns. In dimensional analysis, v is an intermediate variable, the introduction of which reduces the above 4-parametric Π space to a 3-parametric one. For this purpose, the 4 numbers listed above must be formulated with v instead of with q:

Nusselt number	$Nu \equiv$	$\alpha\,D/\lambda$
Prandtl number	$Pr \equiv$	$c\,\eta/\lambda$
Reynolds number	$Re \equiv$	$v\,D\,\rho/\eta$
Froude number	$Fr^* \equiv$	$v^2\,\rho/(D\,g\Delta\rho)$

and must then be combined in such a way as to eliminate D:

Stanton number: $$St \equiv \frac{Nu}{Re\,Pr} \equiv \frac{\alpha}{v\,c\,\rho}$$

Prandtl number $$Pr \equiv c\,\eta/\lambda$$

$$ReFr^* \equiv \frac{v^3}{\nu\,g}\frac{\rho}{\Delta\rho}$$

This results in:

$\{ St, Pr, ReFr^* \}$

We would have obtained this 3-parametric Π set if we had arranged the relevance list at the beginning according to the finding that v has to replace q and had subjected it to dimensional analysis.

$\{ \alpha;\ \rho, \eta, c, \lambda;\ v, g\Delta\rho \}$

	ρ	η	$g\Delta\rho$	c	α	λ	v
M	1	1	1	0	1	1	0
L	−3	−1	−2	2	0	1	1
T	0	−1	−2	−2	−3	−3	−1
Θ	0	0	0	−1	−1	−1	0

Z_1	1	0	0	0	1/3	0	−2/3
Z_2	0	1	0	0	1/3	1	1/3
Z_3	0	0	1	0	1/3	0	1/3
Z_4	0	0	0	1	1	1	0

$Z_1 = M + T + A - 4/3\ \Theta$ $\qquad$ $Z_2 = 3M + L + T + A + 2/3\ \Theta$

$Z_3 = A + 2/3\ \Theta$ $\qquad$ $Z_4 = -\Theta$

$$A = -1/3\,(3M + L + 2T)$$

$$\Pi_1 \equiv \frac{\alpha}{(\rho\ \eta\ g\Delta\rho)^{1/3}\ c} \qquad \Pi_2 \equiv \lambda/\eta\ c \qquad \Pi_3 \equiv \frac{v\ \rho^{2/3}}{(\eta\ g\Delta\rho)^{1/3}}$$

$$\Pi_1\,\Pi_3^{-1} \equiv \frac{Nu}{Re\ Pr} \equiv St \qquad \Pi_2 \equiv Pr^{-1} \qquad \Pi_3^{\,3} \equiv ReFr^*$$

The reason why this dimensional analysis is presented here briefly is because it leads to more complicated Π numbers than usual as shown by the structure of the resulting residual matrix!

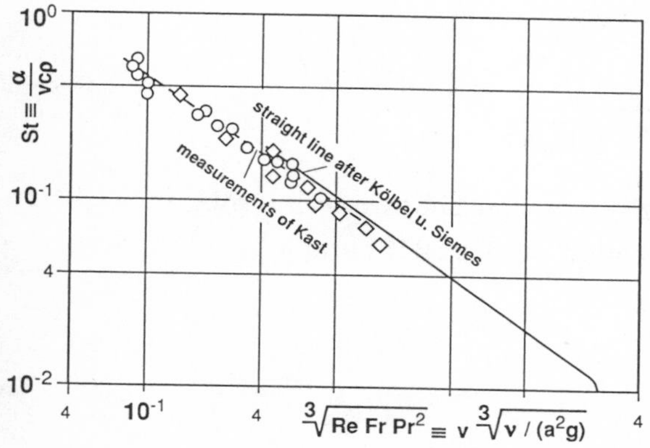

Fig. B 2.1: Heat transfer characteristic of a bubble column. Taken from [B18]. (Thermal diffusivity $a \equiv \lambda/\rho\ c$)

The evaluation of the test results in Fig. B2.1 shows that the 3-dimensional Π space can be further reduced to a 2-dimensional one through product combination $ReFrPr^2$. Apart from using his own measurements, *Kast* also included extensive experimental material supplied by *Kölbel* et al.[24)]

A direct comparison of the heat transfer behaviour of a mixing vessel and a bubble column is only possible if the liquid is aerated in the mixing vessel because then the hydrodynamics is determined in both cases by the same process numbers (Re *and* Fr). This comparison can be effected using our own results (*Zlokarnik* [B19]) for heat transfer in a vessel with hollow stirrer, see Fig. B2.2. It shows that in gas/liquid contacting in mixing vessels approximately twice as much heat can be transferred through the wall as in a bubble column!

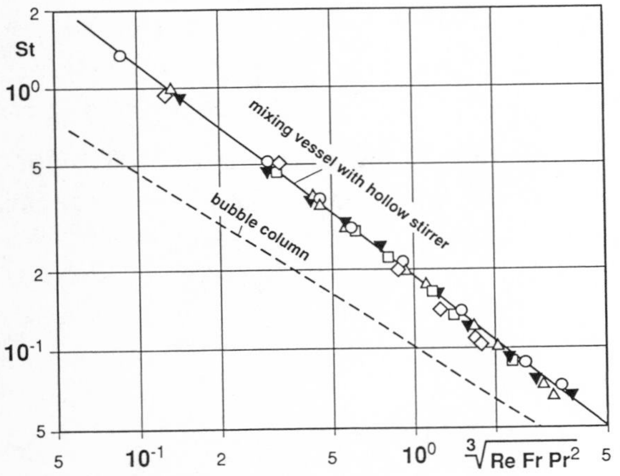

Fig. B2.2: Comparison of the heat transfer behaviour of a mixing vessel with self-aspirating hollow stirrer [B19] and a bubble column [B18].

24) H. Kölbel et al, Chem-Ing.-Tech. **30** (1958) 400/404 and 729/734 as well as **32** (1960), 84/88

Example B3:

Time course of temperature equalization in a liquid with temperature-dependent viscosity in the case of free convection. (Taken from [A13])

Temperature equalization is to be effected in a viscous Newtonian liquid located in a thin-walled copper cylinder. This will be transferred alternately from one thermostat with T_1 into a second with $T_2 > T_1$. Two questions arise:

a) How can the T(t) curve at a selected measuring point (e.g., axis) be described by means of a dimensional analysis ?

b) Do identical T(t) curves result when measured under conditions required for similarity? This question is particularly important if the temperature coefficients of viscosity of the liquids employed differ substantially.

The *target* quantity in this process is the temperature T of the liquid. We will include the cylinder diameter d as characteristic *geometric* parameter in the relevance list. The *physical* properties are the density ρ, the viscosity η, the heat capacity c, the thermal conductivity λ, and the temperature coefficients of viscosity γ and of density ß. The *process* parameters are the experimental time t, gravitational acceleration g (because of the density differences on account of the temperature field) and the two characteristic temperatures:

$T_o = (T_1+T_2)/2$,

to which the numerical values of all physical properties are related and

$\Delta T = T_2 - T_1$

as the maximum temperature difference for the respective temperature equalization.

This leads to the following 12-parametric relevance list:

$\{T;\ d;\ \rho, \eta, c, \lambda, \gamma, \beta;\ g, T_o, \Delta T, t\}$

from which eight dimensionless numbers will be obtained in conjunction with the dimensional system { M, L, T, Θ}. One of these numbers will represent the energy dissipation which cannot be of relevance in this process. In the interest of transparency it is advisable to use the dimensional system which has been extended to include the quantity of heat {M, L, T, Θ, H}, as already demonstrated in Example B1, because in this case the mechanical equivalent of heat J must be incorporated and this identifies the superfluous dimensionless number.

	ρ	d	η	ΔT	c	t	g	λ	J	γ	β	T	T_o
M	1	0	1	0	−1	0	0	0	1	0	0	0	0
L	−3	1	−1	0	0	0	1	−1	2	0	0	0	0
T	0	0	−1	0	0	1	−2	−1	−2	0	0	0	0
Q	0	0	0	1	−1	0	0	−1	0	−1	−1	1	1
H	0	0	0	0	1	0	0	1	−1	0	0	0	0
Z_1	1	0	0	0	0	1	−2	0	−2	0	0	0	0
Z_2	0	1	0	0	0	2	−3	0	−2	0	0	0	0
Z_3	0	0	1	0	0	−1	2	1	2	0	0	0	0
Z_4	0	0	0	1	0	0	0	0	−1	−1	−1	1	1
Z_5	0	0	0	0	1	0	0	1	−1	0	0	0	0

In addition to the obvious dimensionless numbers

$\gamma\Delta T$, $\beta\Delta T$ (resp. β/γ), $T/\Delta T$ und $T_o/\Delta T$

the following four dimensionless numbers appear:

$$\Pi_1 \equiv \frac{t\,\eta}{\rho\, d^2} \qquad \Pi_2 \equiv \frac{g\,\rho^2\, d^3}{\eta^2} = \frac{g\, d^3}{\nu^2} \equiv \mathrm{Ga}$$

$$\Pi_3 \equiv \frac{\lambda}{\eta\, c} \equiv \mathrm{Pr}^{-1} \qquad \Pi_4 \equiv J\,\rho^2\, d^2\, \Delta T\, c\,/\,\eta^2$$

This 8-parametric Π set can be streamlined by the following physically founded considerations:

1. Π_4 is irrelevant because the dissipated energy is negligibly small in the case of free convection.

2. $T_o/\Delta T$ is superfluous because the material function $\eta(T)$ can be represented invariantly, see Sect. 4.2 on p. 54.

3. The gravitational acceleration g and hence the Ga number only occurs together with $\gamma\Delta T$ as the Grashof number $\mathrm{Gr} \equiv \gamma\Delta T\,\mathrm{Ga}$ (see p.54).

4. In the case of *creeping* motion, it is necessary (a) to transform the mass-related heat capacity c in Π_3 into a volume-related one ($c\rho$) and (b) g and ρ can only occur as gravity $g\rho$.

These requirements have been fulfilled when Gr has been multiplied with Pr and the process number Π_1 has been multiplied with Π_3 ($= Pr^{-1}$) to give the Fourier number Fo:

$$\Pi_1 \Pi_3 = \frac{t\lambda}{\rho c\, d^2} \equiv Fo.$$

Finally, once the target number $T/\Delta T$ has been replaced by a temperature number

$$\Theta \equiv \frac{T}{\Delta T} - \frac{T_o}{\Delta T} + \frac{1}{2} = \frac{T - T_1}{\Delta T}$$

which is standardized to one, the complete Π set results

$$\boxed{\{\Theta, Fo, GrPr, \gamma\Delta T, Pr, \beta/\gamma\}}$$

The following remarks apply to the influential range of individual dimensionless numbers:

1) Free convection does not come into play at very small values of Gr. Then temperature equalization takes place as in a solid body according to the process equation:
$\Theta = f(Fo)$.

2) In the range of creeping flow as a result of free convection, the process equation
$\Theta = f(Fo, GrPr)$
is applicable for small values of $\gamma\Delta T$.

3) The following applies only for larger values of $\gamma\Delta T$:
$\Theta = f(Fo, GrPr, \gamma\Delta T)$.

In order to verify these facts, measurements were performed in 3 geometrically similar copper cylinders with D = 30.0; 37.8 and 47.2 mm using 5 different liquids (glycerol, superheated steam cylinder oil, silicone oil, Baysilon M 1000, Desmophen 1100 and HD oil SAE 90). It is important to point out that the standard representation of their material functions $\eta(T)$ is almost consistent.

Fig. B3.1 shows the Θ(Fo) curves for heating and cooling of superheated steam cylinder oil at constant values of GrPr but two different $\gamma\Delta T$ values. That the non-constancy of $\gamma\Delta T$ is responsible for the fact that the curves do not correlate is documented by Fig. B3.2. This shows the temperature curves of four different liquids which have different Pr and β/γ values but were measured at GrPr = idem and $\gamma\Delta T$ = idem.

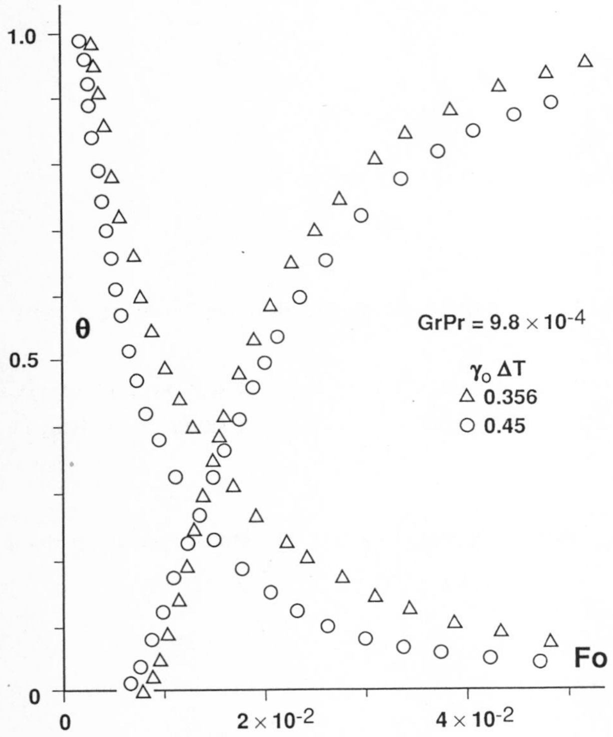

Fig. B3.1: Θ(Fo) curves for cooling and heating of a single liquid (superheated steam cylinder oil). Taken from [A13].

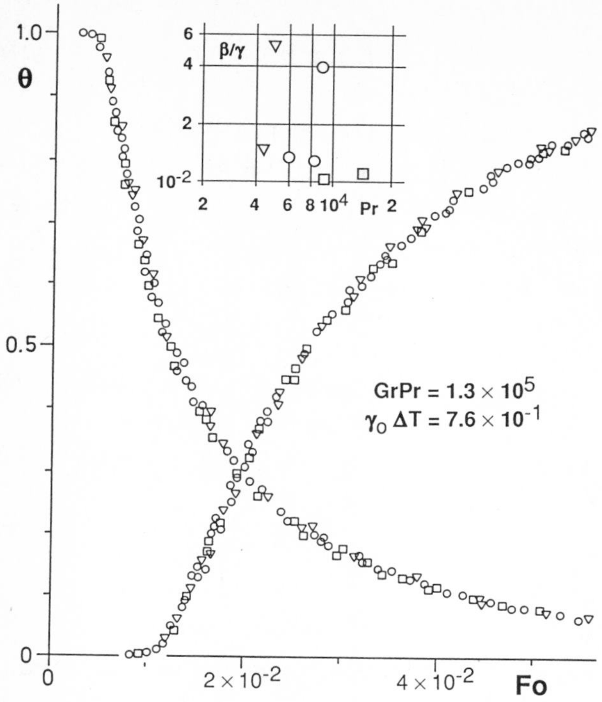

Fig. B3.2: Θ(Fo) curves for cooling and heating of four different liquids at GrPr and $\gamma \Delta T$ = idem but Pr, $\beta/\gamma \neq$ idem. Taken from [A13].

Example B4:

Mass transfer in the gas/liquid system in mixing vessels (bulk aeration) and in biological waste water treatment pools (surface aeration)

The mass transfer process (absorption, desorption: "sorption") in gas/liquid contacting is described with respect to the so-called two-film theory[25] (c.f. Fig.C4.1) with the overall mass transfer equation:

$$\boxed{G = k_L \, A \, \Delta c}$$

---

25) Laminar layers with the thickness δ form on both sides of the inter face. Mass transfer through these is only effected per diffusion. k is defined as $k \equiv \mathfrak{D}/\delta$. The gas-side mass transfer resistance is negligible in comparison to the liquid-side resistance: $k_G >> k_L$.

where:

G - mass transfer rate through the interface [kg s^{-1}]

k_L - liquid-side mass transfer coefficient [m s^{-1}]

A - interfacial area (sum of the surfaces of all gas bubbles) [m^2]

Δc - characteristic concentration difference [kg m^{-3}] of the dissolved gas between the interface and the liquid bulk; in the case of surface aeration (c_S is not a field variable) this difference is given by $\Delta c = c_S - c$; it is assumed that the interface is always saturated with the gas to be dissolved (c_S) according to the Henry´s law.

While the above mass transfer equation (including the definition of Δc) gives a good description of the conditions for surface aeration, uniform distribution of gas bubbles in the liquid is assumed for bulk aeration and mass transfer is formulated in relation to volume:

$$\boxed{G/V = k_L\,(A/V)\,\Delta c = k_L\,a\,\Delta c}\ .$$

Since both k_L and the volume-related interfacial area $a \equiv A/V$ are not easily accessible for measurement, they are combined to form the overall mass transfer coefficient k_La which is then defined with the above mass transfer equation:

$$\boxed{k_La \equiv \frac{G}{V\,\Delta c}}$$

Since part of the gas is absorbed as it bubbles through the liquid column, the composition of the gas mixture is changed. Furthermore, the pressure in the liquid is higher at the gas inlet than in the head space above it. This difference in partial pressure is taken into account by the mean logarithmic concentration difference Δc_m:

$$\Delta c_m = \frac{c_1 - c_2}{\ln\left(\dfrac{c_1 - c}{c_2 - c}\right)}$$

c_1 and c_2 are the saturation concentration under the p-, T-, x- conditions of gas inlet ($_1$) and gas outlet ($_2$); c is the concentration of the gas dissolved in the liquid bulk. (x - mol fraction of the transfer component of the gas mixture).

Selection of the volume-related and hence intensively formulated variable $k_L a$ as *target* quantity of the mass transfer process implies the following consequences:

a) Since a quasi-uniform material system is assumed, $k_L a$ should not depend on *geometric* parameters.

b) On account of $k_G \gg k_L$, $k_L a$ must be independent of the *physical* properties of the gas phase.

c) Since the target quantity $k_L a$ is an intensity variable, the *process* parameters must also be formulated intensively.

According to these premises, the relevance list must be formed with the following parameters: *Target quantity*: $k_L a$; *physical* properties: Density ρ, viscosity η, diffusivity $\mathcal{D}$ and the coalescence parameters S_i of the liquid phase. Despite extensive research, the coalescence phenomena still have not been clarified to such an extent as to permit explicit formulation of the coalescence parameters (see Example A2 on p. 69)! *Process* parameters: Volume-related mixing power P/V, superficial velocity v of the gas and gravitational acceleration g [26]. (The decision in favour of P/V and v instead of P/q and q/V was based on extensive research results [27] obtained in the last two decades.)eee

$$\{ k_L a; \rho, \eta, \mathcal{D}, S_i; P/V, v, g \}$$

	ρ	η	g	$k_L a$	P/V	v	$\mathcal{D}$
M	1	1	0	0	1	0	0
L	−3	−1	1	0	−1	1	2
T	0	−1	−2	−1	−3	−1	−1
Z_1	1	0	0	1/3	2/3	- 1/3	−1
Z_2	0	1	0	− 1/3	1/3	1/3	1
Z_3	0	0	1	− 2/3	4/3	1/3	0

26) $g\Delta\rho = g(\rho - \rho')$ should have been written here but $g\rho$ would have resulted on account of $\rho \gg \rho'$. This would lead to $g\rho/\rho = g$ in the dimensionless number, see Example A2, p.69.

27) K.van't Riet, Ind. Eng. Chem. Process Des. Devel. **18** (1979) 3, 357

$Z_1 = M+T+2A; \qquad Z_2 = 3M+L+T+A \qquad Z_3 \equiv A = -1/3\,(3M+L+2T)$

$$\Pi_1 \equiv k_L a \left(\frac{\eta}{\rho\, g^2}\right)^{1/3} = k_L a \left(\frac{\nu}{g^2}\right)^{1/3} \equiv (k_L a)^*$$

$$\Pi_2 \equiv \frac{P/V}{(\rho^2\, \eta\, g^4)^{1/3}} = \frac{P/V}{\rho\, (\nu\, g^4)^{1/3}} \equiv (P/V)^*$$

$$\Pi_3 \equiv \frac{v\, \rho}{(\eta\, g)^{1/3}} = \frac{v}{(\nu\, g)^{1/3}} \equiv v^* \qquad \Pi_4 = \frac{\mathfrak{D}\, \rho}{\eta} = \frac{\mathfrak{D}}{\nu} \equiv Sc^{-1}$$

The following Π set resulted here:

$\{(k_L a)^*, (P/V)^*, v^*, Sc, Si^*\}$

Fig. B4.1 shows a correlation [B19] of mass transfer measurements in this Π space under non-steady-state conditions. The measurements were performed by several authors in the water/air system which is very prone to coalescence, using the turbine stirrer (see Example A2) as mixing device. The measurements cover an experimental scale of $\mu = 1 - 100$!

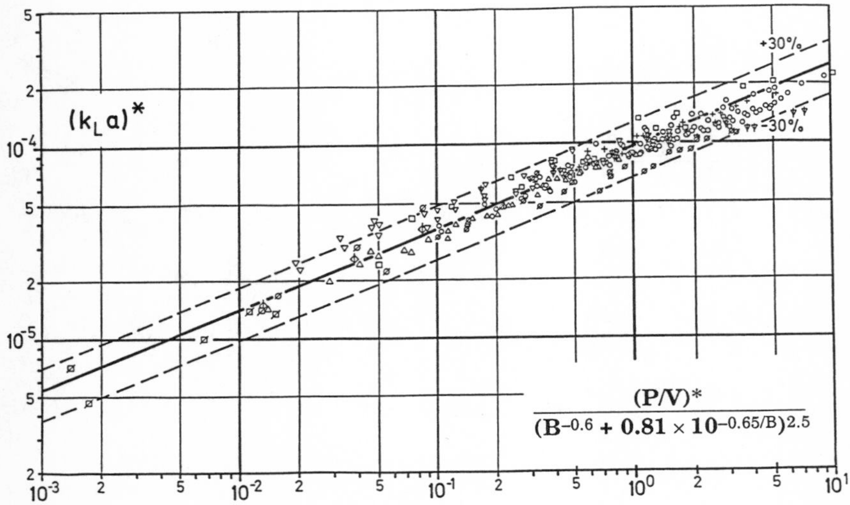

Fig. B4.1: Sorption characteristic of a mixing vessel with turbine stirrer for a coalescing material system (water/air). Taken from [B19]. (**B** stands here for v*)

Fig. B4.2 in contrast shows results of mass transfer in the system aq 1n sodium sulphite solution/air which were obtained under steady-state conditions in vessels with hollow stirrers on the scale $\mu = 1 - 5$ [B19/20]. In this material system, the high salt concentration (70 g/l) causes extreme suppression of gas bubble coalescence. In the case of the self-aspirating hollow stirrer (see Example A3), the stirrer power and gas throughput are coupled via the rotational speed and are therefore dependent on each other. Consequently v* does not occur explicitly in the representation in Fig. B4.2 because it is a function of (P/V)*.

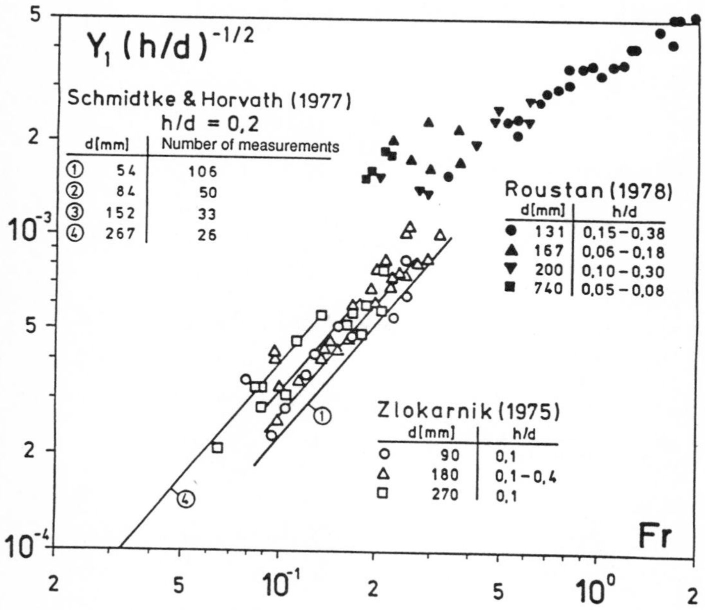

Fig. B4.2: Sorption characteristic of a mixing vessel with self-aspirating hollow stirrer and a material system (70 g Na_2SO_3/l) with strongly suppressed coalescence . Taken from [B17/18].

In the case of surface aeration, mass transfer essentially takes place through the liquid *surface* which has been agitated by the stirrer. However, since the change in concentration is measured in the liquid volume: ln c = f(t), the volume-related k_La results here too. This must be converted into the target quantity $k_LA = k_LaV$ which is applicable in this case through multiplication with the liquid volume V of the pool.

k_LA is not an intensively formulated quantity. Therefore the restrictions applied in connection with k_La are not valid here. The relevance list includes the aerator diameter d (the pool diameter D must be much larger than the aerator diameter d, in which case D/d is also irrelevant!), the physical properties of the liquid ρ, ν and $\mathcal{D}$ and the process parameters rotational speed of the stirrer n and gravitational acceleration g (the latter is particularly relevant because it decisively influences the development of the spray ring and waves). The coalescence parameters S_i in contrast have no influence in this case. The following is then obtained:

$$\{ k_LA; d; \rho, \eta, \mathcal{D}; n, g \}$$

This results in four dimensionless numbers

$$(k_LA)^* \equiv \frac{G}{d^3 \Delta c}\left(\frac{\nu}{g^2}\right)^{1/3}; \quad Fr \equiv \frac{n^2 d}{g}; \quad Sc \equiv \frac{\nu}{\mathcal{D}}; \quad Re \equiv \frac{n\, d^2}{\nu} \quad \text{resp.} \; Ga \equiv \frac{Re^2}{Fr} \equiv \frac{g\, d^3}{\nu^2}$$

Since the design and scale-up data for surface aerators of waste water are of interest, the test liquid cannot be selected at random (similar to Froude´s problem with the ship, p. 42); In tests with tap water, Ga can be varied only by means of a scale alteration.

Fig. B4.3 shows the results of three research studies [B23] in which a turbine stirrer (see sketch in Fig. A2.1) positioned in the water surface was employed as model surface aerator. It is apparent that the Fr number is extremely important. Only the extensive measurements by *Schmidtke and Horvath* which were performed on the scale μ = 1:5 and statistically evaluated (see drawn straight lines in Fig. B4.3) indicate with certainty the slight influence of Ga. The process characteristic for this stirrer at h/d = 0.2 (h - immersion depth of the stirrer blades) is:

$$(k_LA)^* = 1.41 \times 10^{-4}\, Fr^{1.205}\, Ga^{0.115}$$

Surface aerators for biological waste water treatment plants have always been *incorrectly* designed and scaled-up on the basis of the volume(!)-rela-

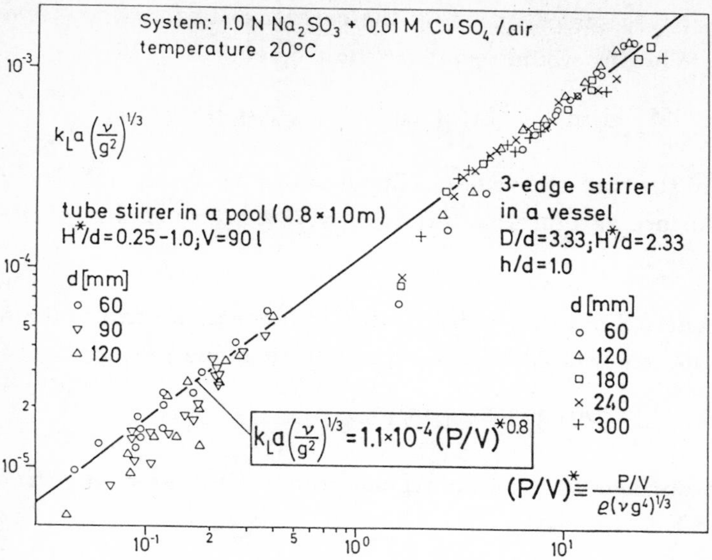

Fig. B4.3: Sorption characteristic of the turbine stirrer as surface aerator. Taken from [B23]. Y_1 stands here for $(k_L a)^*$.

ted stirrer power P/V. (Since the pool depth is kept constant at $H \approx 4$ m independent of pool size, this rule can also be interpreted as P/D^2 = const and for D/d = const as P/d^2 = const). Since model measurements have shown [B 22] that the power characteristic of a surface aerator is given by Ne(Fr), we can check this scale-up criterion using the above sorption characteristic.

Starting with the two characteristics:

$$(k_L A)^* \, Ga^{-0.115} = f_1(Fr) \qquad \text{and} \qquad Ne = f_2(Fr)$$

we will first form a combination of numbers which is a dimensionless expression of the efficiency $E \equiv G/P$ [kgO_2/kWh] of the surface aerator and which itself - since all arguments of the above characteristics are functions of Fr - is also a function of Fr:

$$\boxed{\frac{(k_L A)^* \, Ga^{-0.115}}{Ne \, Fr^{1.5}} \equiv \frac{G \, d^{0.155}}{P \, \Delta c} \, \phi(\rho,\nu,g) = f_3(Fr)}$$

If a surface aerator was scaled-up using the correct criterion *Fr = idem* for Δc = const, the following would result for the efficiency E = G/P:

$$Fr = idem \;\rightarrow\; E\, d^{0.155} = idem \;\rightarrow\; E_H = E_M\, \mu^{0.155} \qquad (\mu \equiv d_M/d_H)$$

This means that, in the case of a scale-up factor of μ = 5 (10), the efficiency of the prototype surface aerator amounts to only 78% (70%) of that of the model!

If the scale-up is carried out on the basis of $P/V = const$, we can make a statement on the volume-related oxygen input G/V. It follows that:

$$(G/P)\, d^{0.155} = idem \;\rightarrow\; (G/V)\, d^{0.155} \sim (P/V)$$

This means that surface aerator enlargement has the same *negative* effect on G/V as on E = G/P.

Example B5:

Design and scale-up of injectors as gas distributors in bubble columns

Injectors are two-component nozzles which utilize the kinetic energy of the liquid propulsion jet to disperse the gas continuum into very fine gas bubbles[28] and to distribute them in the liquid. Their *advantage* over stirrers is that the liquid jet causes gas dispersion directly while the stirrer has to set the entire contents of the vessel in motion in order to generate the necessary shear rate in the liquid. Their *disadvantage* is the predominance of severe coalescence on account of the high gas bubble density in the free jet of the gas/liquid dispersion. However, in contrast to stirrers, the injector cannot cause redispersion of the large gas bubbles.

Design fundamentals for the so-called slot injector – see Fig. B 5.1 – are presented in the following; The shape of its mixing chamber performs two different functions:

a) Due to the converging walls of the casing, the shear rate of the propulsion jet increases along the mixing chamber; however, because the cross-

[28] In contrast to injectors, ejectors are two-component nozzles in which the kinetic energy of the liquid jet is used to generate suction.

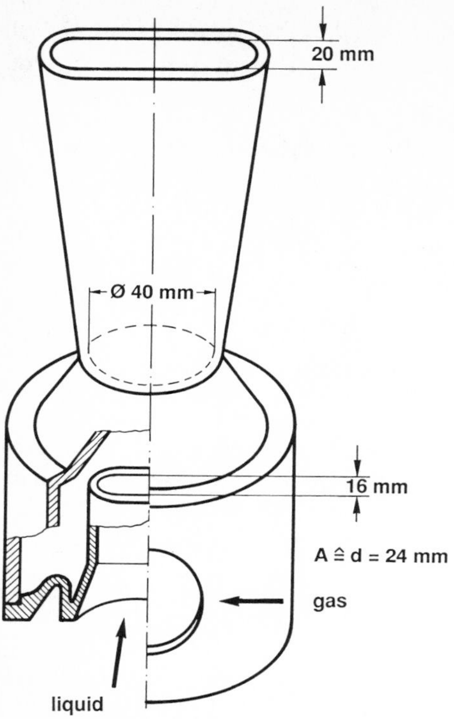

Fig. B 5.1 : Sketch of the slot shaped injector

sectional area of the mixing chamber remains unchanged, this does not result in an additional pressure drop.

b) The free jet of the gas/liquid dispersion leaves the slot-shaped mouthpiece in the form of a ribbon which mixes more quickly into the surrounding liquid than a jet with circular cross-section. This counteracts bubble coalescence.

For optimum design of injectors for gas/liquid contacting their pressure drop and sorption characteristics must be known. The latter is required to establish the necessary gas and liquid flowrate; the former is needed to dimension the conveying devices (pumps and blowers).

a) The pressure drop characteristics of an injector are based on the following relevance lists:

for the *gas throughput* : $\{\Delta p; d_M; \rho, \nu_L; q, q_L\}$
for the *liquid throughput* : $\{\Delta p_L; d, \rho_L, \nu_L; q, q_L\}$

The meanings are as follows: Δp - pressure drop of the respective medium in the propulsion jet nozzle of diameter d resp. in the mixing chamber of diameter d_M; q - throughputs; ρ and ν densities resp. kinematic viscosities of the respective medium. (Gas: without index, liquid: index L). The following Π sets result:

for the *gas throughput* : $\{ Eu \equiv \frac{\Delta p\, d_M^4}{\rho\, q^2} \;;\; q/q_L \;;\; Re \equiv q_L/(\nu_L\, d_M) \}$

for the *liquid throughput* : $\{ Eu_L \equiv \frac{\Delta p_L\, d^4}{\rho_L\, q_L^2} \;;\; q/q_L ;\; Re_L \equiv q_L/(\nu_L\, d) \}$

Measurements on different injector designs have shown that Re is irrelevant in the range of Re > 10^4. Therefore q/q_L is the only process number in both cases; see Fig. B 5.2.

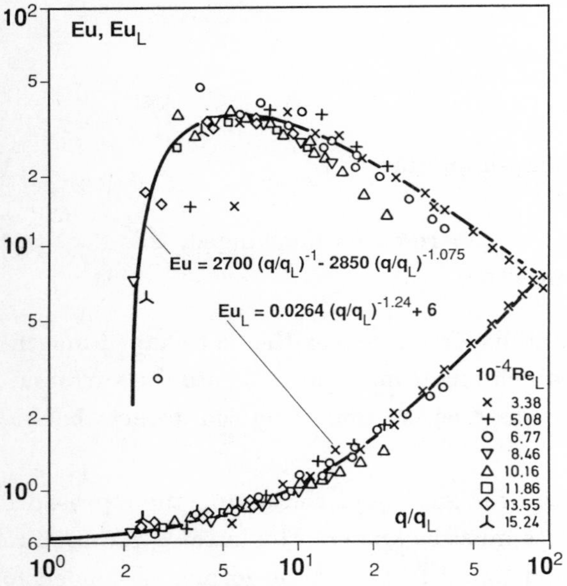

Fig. B 5.2: Pressure drop characteristics of an injector; q is related to standard conditions

b) To follow the sorption characteristic of a mixing vessel, the sorption characteristic of an injector will also be formed with intensively formulated process parameters, because the target quantity k_La is also an intensity quantity. Instead of the gas throughput q, we will introduce its superficial velocity $v \sim q/D^2$, since it has proved suitable for the correlation of k_La values in laboratory bubble columns: k_La/v = const. (see remarks on p.49). Instead of the liquid throughput q_L, we will use the power of the liquid jet $P_L = \Delta p_L q_L$, related to the gas throughput q: P_L/q [B8]. Including the physical parameters as in Example B4, the following relevance list results:

$$\{ k_La/v; \rho, \nu, \mathcal{D}, S_i; P_L/q, g \},$$

This leads to the following Π set:

$$\{(k_La/v)^*, Sc, S_i^*, (P_L/q)^*\}$$

Sorption number $\qquad Y \equiv (k_La/v)^* \equiv \frac{k_La}{v}\left(\frac{\nu^2}{g}\right)^{1/3}$

Dispersion number $\qquad X \equiv (P_L/q)^* \equiv \frac{P_L/q}{\rho(\nu g)^{2/3}}$

Fig. B 5.3 shows the sorption characteristics of the industrial-size slot injector which were measured in a bubble column of $3^{\varnothing} \times 8$ m in dependence on the common salt (NaCl) content to purposefully influence the coalescence behaviour of the material system. The following shows why

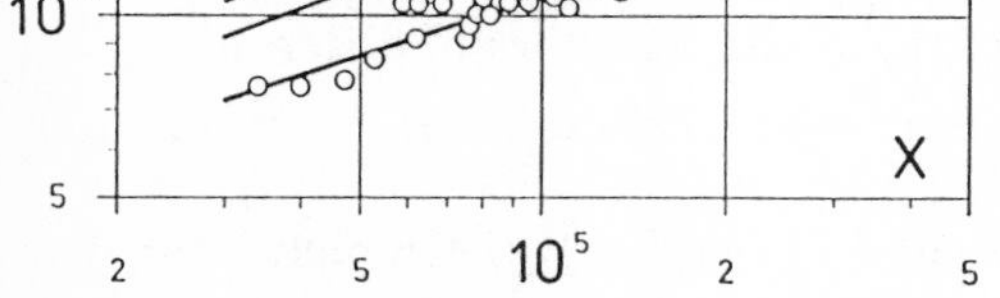

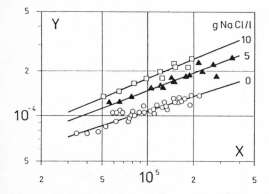

Fig. B 5.3: Sorption characteristics of the industrial-size slot injector in dependence on the degree of coalescence of the system. Taken from [B8 b].

these characteristics can be used only as design data for optimizing the running conditions of this *particular* injector and why they are certainly not suitable for a scale-up of this device.

In fact, the concept of the quasi-homogeneous gas/liquid mixture, which was verified in bubble columns with perforated plates as gas distributors, proves to be *totally inappropriate* when injectors are used as gas dispersers. The explanation for this fact is based on two main aspects of injectors and demonstrates that, when they are used, coalescence takes place both in the free jet and in the aeration tank, while in the case of gas distribution with perforated plates this process has already been completed just above the perforated plate. (In other words: There is no justification for formulating the target number with intensity variables which lead to $Y \equiv (k_L a/v)^*$).

1.The dispersing effect of the liquid propulsion jet is restricted to its circumference which, in case of geometrically similar scale-up, increases linearly ($u = \pi d$) while its cross-section increases quadratically ($A = \pi\ d^2/4$). This means that with increasing diameter of the device, an increasingly smaller fraction of the liquid throughput is available for dispersion: *The dispersion efficiency of injectors is inversely proportional to the scale!*

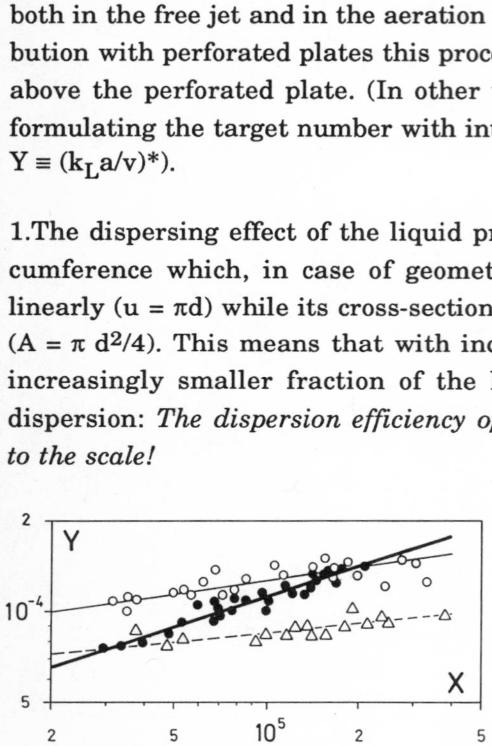

Fig. B 5.4: Sorption characteristics of three slot injectors of different shape and size (explanation in text). Taken from [B8 a].

This is confirmed by the results presented in Fig. B 5.4 which gives the Y(X) correlation for two geometrically similar slot injectors in the scale μ = 1:2. The way out of this dilemma was to avoid scaling-up under geometrically similar conditions and to increase only the diameters and not the lengths by a factor of 2. As a result, all angles are doubled: the shear rates are enhanced and the free jet fans out more which further suppresses coalescence.

2. The Π set for the sorption characteristic only takes account of gas dispersion and not of bubble coalescence in the liquid. According to the laws governing the free jet, the propulsion jet of the gas/liquid dispersion sucks in the liquid surrounding it, whereby it loses its kinetic energy and decomposes into a gas bubble swarm. This process is extremely scale dependent and would need to be considered after inclusion of all relevant geometric parameters.

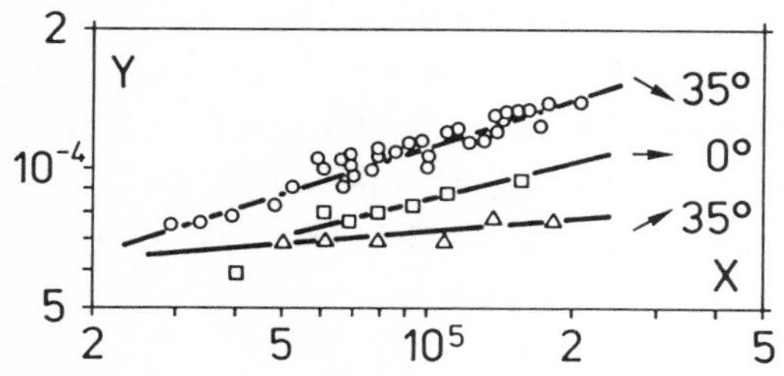

Fig. B 5.5: Influence of the angle of inclination of the free jet on bubble coalescence and hence on Y(X). Taken from [B8 b].

To give an example of the dramatic influence which the geometric parameters can have on coalescence behaviour, Fig. B 5.5 shows Y(X) correlations for the industrial-size slot injector which were obtained in a vessel of 3∅ × 8 m (water height). The injector was positioned 1 m above the bottom at the vessel wall in such a way that its axis formed an angle of 0°, + 35° resp. – 35° with the horizontal. In the latter case, the free jet was pointed towards the bottom and decomposed into the bubble swarm just above it. Near the bottom, the suction of the free jet is weakest on account of bottom friction; furthermore, the bubble swarm which has formed does not exert a strong "chimney effect" there. Consequently liquid entrainment into the free jet is suppressed at exactly that point at which it would be particularly supportive of coalescence on account of the weakened kinetic energy of the free jet.

The sorption characteristics Y(X) presented in Figures B 5.2–4 can therefore only be used as design data for the injector type measured in the respective case. They are not to be used for scale-up!

In view of the fact that injectors exhibit favorable characteristics for gas/liquid contacting which make them superior to stirrers, systematic research in this area would be most desirable. However, the experiments would have to be performed in differently sized equipment (bubble columns with injectors).

Example B6:

Scale-up problems relating to continuous, carrier-free electrophoresis [*)]

Electrophoresis makes use of differences in electrophoretic mobility of electrically charged particles (biomolecules, micro-organisms etc.) in the homogeneous, rectified electrical field for their separation. Thanks to the excellent resolution and mild operating conditions, this is currently the best analytical method for protein separation, purification and characterization. It is also used as a preparative separation method in the range of some grammes per hour.[29)]

In the case of *carrier* electrophoresis, a sheet of paper, starch, polyacrylamide or agarose gel or something similar, which is saturated with buffer solution, is placed between the electrodes which generate a homogeneous, rectified electrical field. The mixture of substances which is supplied at one point, is resolved into its individual components in the course of time. Carrier electrophoresis is therefore a *discontinuous* separation process.

Carrier-free, *continuous* electrophoresis is utilized for preparative work. In this process, the two electrodes form a parallel, plane slot or annulus of some mm to some cm in width through which the buffer solution migrates. The material mixture to be fractionated is injected into the carrier

---

*) I am grateful to *Dr. Th. Melin* who introduced me to this problem

29) See overviews such as: C. I. Ivory, Sep.Sci.Tech. **23** (1988) 8&9, 876/912 or R. Rutte, Swiss Chem **11** (1989) 10, 33/37

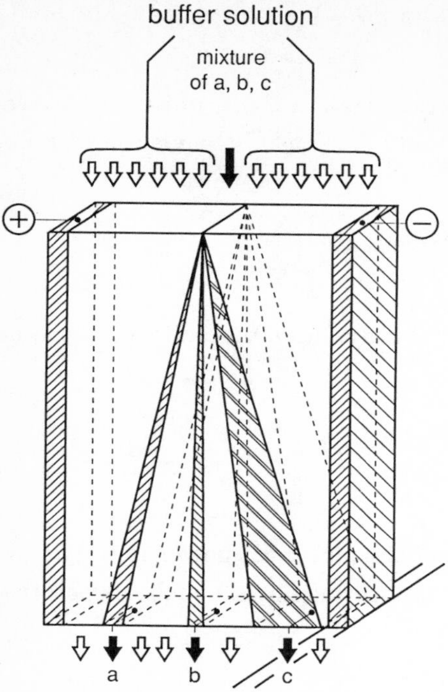

Fig. B 6.1: Sketch of the carrier–free electrophoresis cell

solution at one point in this slot and, after its resolution into separate fractions, is drawn off with the buffer solution at many points along the slot, see Fig. B 6.1.

Before compiling the relevance list for this process, it must be considered in greater depth:

The resolution is characterized by a bandwidth b (which is as small as possible) and a clearance s between the bands (which is as large as pos sible). Both parameters are dependent on the same influencing variables.

The *resolution effect* is therefore defined with the quotient s/b and is taken as the *target number* of the process.

One of the main material-related parameters will be *electrophoretic mobility* μ. It is defined as the velocity v of the particles in relation to the electrical field strength E. It results from the balance between the electrical field force F_{el} and the frictional force F_r.

The following is valid for the electrical field force F_{el}:

$$\boxed{F_{el} = e\,z\,E}$$

where e [Coulomb = A s] is the electrical charge, z the number of charge carriers per particle (e.g., biomolecules) and E [V m^{-1}] the electrical field strength.

The following is valid for the frictional force F_r:

$$\boxed{F_r = 6\,\pi\,r\,\eta\,v}\;;$$

where r is the particle diameter and η the dynamic viscosity of the medium.

The following results for electrophoretic mobility:

$$\boxed{\mu = v/E = e\,z\,/(6\,\pi\,r\,\eta\,)}$$

Furthermore, it is important to know that the electrical power is fully converted into heat. This causes convective material flows which oppose the fractionation process.

We are going to consider two different operating modes. In the first, a *creeping flow* predominates between two plane-parallel electrodes. The electrodes are not cooled in this case; consequently $\Delta\vartheta$ is not a freely selectable process parameter. In the second operating mode, in contrast, the operating conditions are those of the so-called *Biostream separator*.

Case 1: *Creeping flow*

This boundary condition has the following consequences for the relevance list for the problem:

1. The density ρ as such does not play any role;
2. The specific heat c_p must be formulated as volume-related: ρc_p ;
3. The temperature coefficient β of density only acts in conjunction with gravitational acceleration g and must be combined with the density: $\rho g \beta$

The influencing variables of this process are therefore:

1. *Geometric* parameters: Characteristic length of the channel l
2. *Physical properties*: The dynamic viscosity η of the buffer solution; the

diffusion coefficient $\mathfrak{D}$ of the substances to be separated in the buffer solution; $\rho g \beta$, the specific heat ρc_p and the thermal conductivity λ. (These three quantities are the only ones which contain temperature in their dimensions; therefore we will formulate them as quotients to obtain $\lambda/\rho g \beta$ and the thermal diffusivity $a \equiv \lambda/\rho c_p$). The electrical physical properties are electrical conductivity λ_{el} (= current density / electrical field strength) and electrophoretic mobility μ.

3. *Process-related* parameters: Volumetric flowrate q and electrical tension $U \sim E\,l$, because this, in contrast to E, is a directly adjustable process variable.

The relevance list of the problem is therefore:

$$\{s/b;\ l;\ \eta,\ \mathfrak{D},\ \lambda/\rho g\beta,\ a,\ \lambda_{el},\ \mu;\ q,\ U\}$$

Without the target number s/b, the dimensional matrix is:

	η	l	$\mathfrak{D}$	μ	$\lambda/\rho g\beta$	a	λ_{el}	q	U	
M	1	0	0	−1	0	0	−1	0	1	
L	−1	1	2	0	3	2	−3	3	2	
T	−1	0	−1	2	−1	−1	3	−1	−3	
I	0	0	0	1	0	0	2	0	−1	
Z_1	1	0	0	0	0	0	1	0	0	$Z_1 = M + I$
Z_2	0	1	0	0	1	0	−2	1	0	$Z_2 = 3M + L + 2T - I$
Z_3	0	0	1	0	1	1	0	1	1	$Z_3 = -M - T + I$
Z_4	0	0	0	1	0	0	2	0	−1	$Z_4 = I$

It leads to the following Π set:

$$\Pi_1 \equiv \frac{\lambda}{\rho g\beta\,\mathfrak{D}\,l};\quad \Pi_2 \equiv \frac{a}{\mathfrak{D}} \equiv \mathrm{Le};\quad \Pi_3 \equiv \frac{\lambda_{el}\,l^2}{\eta\,\mu^2};\quad \Pi_4 \equiv \frac{q}{l\,\mathfrak{D}} \equiv \mathrm{Bo};\quad \Pi_5 \equiv \frac{U\mu}{\mathfrak{D}}.$$

$\Pi_4 \equiv \mathrm{Bo}$ and Π_5 are the two process numbers.

$\Pi_2 \equiv \mathrm{Le}$ (Lewis number) and $\Pi_1\,\Pi_3^{0{,}5} = \dfrac{\lambda}{\rho g\beta\,\mathfrak{D}\,\mu}\left(\dfrac{\lambda_{el}}{\eta}\right)^{0.5}$

are the two material numbers, while the combination of numbers

$$\Pi_1^{-1}\,\Pi_3 = \frac{\rho g \beta\, \mathfrak{D}\, l^3}{\eta\, \mu^2}\,\frac{\lambda_{el}}{\lambda} \equiv \text{"Gr"}$$

provides a modified Grashof number (see p.54).

Taking account of these alterations and the addition of the target number s/b, the above Π set is as follows:

$$\frac{s}{b}\,;\ \frac{q}{l\,\mathfrak{D}} \equiv \mathrm{Bo}\,;\ \frac{U\,\mu}{\mathfrak{D}}\,;\ \frac{\rho g \beta\, \mathfrak{D}\, l^3}{\eta\, \mu^2}\,\frac{\lambda_{el}}{\lambda} \equiv \text{"Gr"};\ \frac{\lambda}{\rho g \beta\, \mathfrak{D}\, \mu}\left(\frac{\lambda_{el}}{\eta}\right)^{0.5};\ \frac{a}{\mathfrak{D}} \equiv \mathrm{Le}.$$

It is apparent that a change in scale while keeping s/b = idem necessitates the retention of *idem* for the mechanical process number Bo and hence $q \sim L$, i.e. an increase in q causes a corresponding increase in scale. However, this condition means that the numerical value of "Gr" is changed very substantially during scale-up. This dimensionless number could only be kept *idem* by changing the physical properties and this naturally cannot be done.

A way out, which has frequently been considered but, understandably, still has not been realized would be to eliminate the influence of "Gr" by exclusion of gravity in the Spacelab. The efforts to suppress smearing of the individual material bands by superimposing a shear flow crosswise to the volume flow in the annulus between the concentric electrodes would seem to be more interesting.

Case 2: The *Biostream Separator*

The "Biostream Separator" operates according to this principle (see footnote 29). The device consists of two concentric cylinders; the inner one is made to rotate in order to generate a Couette flow. We are no longer dealing with a creeping flow, the liquid density ρ is not negligible. The number of *process* parameters is increased from two to four by the rotational speed n of the inner cylinder and, as a result of the possibility of cooling the outer cylinder wall, by the temperature difference ΔT, in this case as a freely adjustable parameter.

The relevance list which has been extended in comparison to Case 1 is now:

$\{s/b;\ l;\ \eta,\ \mathcal{D},\ \lambda/\rho g\beta,\ a,\ \lambda_{el},\ \mu;\ q,\ U$	$n,\ \lambda\Delta T,\ \rho\}$

The extended dimensional matrix supplies the three additional dimensionless numbers:

$$\Pi_6 \equiv \frac{n\,l^2}{\mathcal{D}}; \qquad \Pi_7 \equiv \frac{\lambda\,\Delta T\,l^2}{\eta\,\mathcal{D}^2} \qquad \Pi_8 \equiv \frac{\rho\,\mathcal{D}}{\eta} \equiv Sc^{-1}$$

To simplify test planning and execution, the complete Π set (i.e., the already obtained numbers Π_1 to Π_5 and the three numbers Π_6 to Π_8 given above) is converted as follows to more sensible process and material numbers:

Target number: $\Pi_9 \equiv s/b$

Process number for n : $\Pi_6\,\Pi_8 \equiv \frac{n\,l^2}{\nu} \equiv Re$ (Reynolds number)

for q : $\Pi_4 \equiv \frac{q}{\mathcal{D}\,l} \equiv Bo$ (Bodenstein number)

for E : $\Pi_5 \equiv \frac{U\,\mu}{\mathcal{D}}$

for $\Delta\vartheta$: $\Pi_1^{-1}\Pi_7\Pi_8 \equiv \frac{g\beta\,\Delta T\,l^3}{\nu^2} \equiv Gr$ (Grashof number)

Material numbers: $\Pi_2 \equiv \frac{a}{\mathcal{D}} \equiv Le$ (Lewis number)

$\Pi_8^{-1} \equiv \frac{\nu}{\mathcal{D}} \equiv Sc$ (Schmidt number)

$\Pi_1\,\Pi_3^{0.5} \equiv \frac{\lambda}{\rho g\beta\,\mathcal{D}\,\mu}\left(\frac{\lambda_{el}}{\eta}\right)^{0.5}$

and $\Pi_8^{-1}\,\Pi_6 \equiv \frac{\eta\,l^2\,n^2}{\lambda\,\Delta T} \equiv Br$ (Brinkman number)

The target number is dependent on four process numbers and three pure material numbers. Since the rotation of the inner cylinder generates a negligible "heat of agitation", the last number, the Brinkman number, can be considered to be irrelevant and deleted (see Examples B1 and B3).

The model tests to determine scale-up rules are necessarily performed with the same material system as that treated in the prototype because the alteration of a physical characteristic, e.g. the viscosity – automatically changes the other physical properties – e.g. electrophoretic mobility.

In the model apparatus it will be possible to adjust the four process numbers independent of each other without any problems. In this way, their influence on s/b will be clarified and optimum process conditions for the desired separation operation will be established. (We would like to assume that, in these tests, it will also be possible to determine the influence of Gr on s/b within certain limits by altering ΔT.)

On the next-larger model scale the process numbers for n, q and E can also be set to idem without any problems since they contain clear instructions as to how the respective process variables have to behave in the case of a change in scale. The adjustment of Gr = *idem*, in contrast, will be problematic, since *doubling* the slot width makes it necessary to set a temperature difference which is *smaller* by a factor of 8 on account of $\Delta T \sim l^{-3}$. As a result, the heat produced in the annulus can no longer be removed.

C Examples from the Field of Chemical Reaction Engineering

Introductory Remarks:

Chemical reactions obey the rules of chemical thermodynamics and chemical reaction kinetics; if they take place slowly and without significant heat of reaction in the homogeneous system ("microkinetics") they are not subject to any rules of the theory of similarity.

However, such a reaction course occurs only very infrequently in chemical reaction engineering. Most chemical reactions take place in heterogeneous material systems (liquid/liquid, liquid/gas, liquid/solid, gas/solid) and generate considerable reaction heat. Consequently the genuine chemical action is accompanied by mass and heat transfer processes ("macrokinetics") which are scale-dependent. The course of such chemical reactions will be similar on a small and large scale if the mass and heat transfer processes are similar and the "chemistry" is the same.

In a continuous reaction process, the actual residence time of the reaction partners in the reactor plays a major role. It is governed by the residence-time distribution characteristic of the reactor which gives information on back-mixing (macromixing) of the throughput. This emphasizes the interaction between chemical reaction and fluid dynamics.

Chemical reactions exist, in which mass and/or heat transfer represent the rate-controlling step. If both transfer processes are *simultaneously* effective, special scale-up problems may arise because they obey different laws; see reaction in a catalytic packed-bed reactor, Example C1.

From the point of view of dimensional analysis, a chemical engineering problem presents itself with the appearance of chemical parameters containing the primary quantity amount of substance **N** in their dimensions. We then have to deal with a 5-parametric dimensional system {**M, L, T, Θ, N**}.

Example C1:
Continuous chemical reaction process in a tubular reactor

Historically speaking, *Gerhard Damköhler* (1908 - 1944) was the first to investigate a chemical process in conjunction with mass and heat transfer using the approach of the theory of similarity [18]. In a purely theoretical way he examined the conditions under which scale-up would be possible in the case of (inevitably) partial similarity and he checked the consequences which would result from such a procedure. However, before analysing his method, we shall discuss this problem from the point of view of dimensional analysis.

1. Homogeneous irreversible reaction of the 1st order

A homogeneous, volume-resistent chemical reaction taking place in a tubular reactor is influenced by the mass and heat transfer processes. The flow condition is described by v, d, l, ρ, η (v - flowrate, d resp. l - diameter resp. length of the tube, ρ resp. η - density resp. viscosity of the fluid). All physical properties are related to the known inlet temperature T_o. In contrast to the continuous reaction in the catalytic packed-bed reactor - see section 2 - it is assumed here that c and T differences in radial direction are negligible (turbulent flow regime).

The chemical reaction and its degree of conversion are characterized by the *inlet* and *outlet* concentrations c_o and c_{out} and by the effective reaction rate constants k_{eff}. (Note that the reaction order – here 1st order – governs the dimension of k_o!) At the temperature field actually predominating in the reactor, the effective reaction rate constants k_{eff} in the reactor will adjust themselves in accordance with Arrhenius´ law:
$k_{eff} = k_o \exp(E/RT)$.

The mass and heat transfer is described by
$\{\mathfrak{D}, c_p, \lambda, c_o\Delta H_R, T_o, \Delta T\}$
($\mathfrak{D}$ - diffusion coefficient, c_p - heat capacity, λ - thermal conductivity, $c_o\Delta H_R$ - heat of reaction produced per unit time and volume, T_o - inlet temperature, ΔT - temperature difference between fluid and tube wall). The complete relevance list is therefore:

{v, d, l, ρ, η, c_o, c_{out}, k_o, E/*R*, 𝔇, c_p, λ, $c_o\Delta H_R$, T_o, ΔT}

Only 9 numbers are formed from these 15 dimensional parameters if the heat H is added to the five primary quantities {M, L, T, Θ, N} as the sixth primary quantity: 15 – 6 = 9. If L/d, c_{out}/c_o, E/RT_o, und $\Delta T/T_o$ are anticipated as trivial numbers, the other 5 dimensionless numbers can be obtained using the following simple dimensional matrix:

	ρ	l	k_o	T_o	$c_o\Delta H_R$	v	η	𝔇	c_p	λ	
M	1	0	0	0	0	0	0	0	–1	0	
L	–3	1	0	0	–3	1	–1	2	0	–1	
T	0	0	–1	0	0	–1	–1	–1	0	–1	
Θ	0	0	0	1	0	0	0	0	–1	–1	
H	0	0	0	0	1	0	0	0	1	1	
Z_1	1	0	0	0	0	0	1	0	–1	0	Z_1 = M
Z_2	0	1	0	0	0	1	2	2	0	2	Z_2 = 3M + L + 3H
Z_3	0	0	1	0	0	1	1	1	0	1	Z_3 = -T
Z_4	0	0	0	1	0	0	0	0	–1	–1	Z_4 = Θ
Z_5	0	0	0	0	1	0	0	0	1	1	Z_5 = H

The following dimensionless numbers result:

$$\Pi_1 \equiv \frac{v}{l\, k_o} = (k_o \tau)^{-1} \text{ (residence time } \tau \equiv l/v \text{ at pipe flow!)}$$

$$\Pi_2 \equiv \frac{\eta}{\rho\, l^2\, k_o} = (k_o \tau\, Re\, l/d)^{-1}$$

$$\Pi_3 \equiv \frac{\mathfrak{D}}{l^2\, k_o} = (k_o \tau\, Re\, Sc\, l/d)^{-1}$$

$$\Pi_4 \equiv \frac{c_p\, \rho\, T_o}{c_o \Delta H_R} = Da^{-1} \text{ (Da - Damköhler number)}$$

$$\Pi_5 \equiv \frac{\lambda\, T_o}{c_o \Delta H_R\, l^2\, k_o} = (k_o \tau\, Re\, Pr\, Da\, l/d)^{-1}$$

The power products formed with the aid of the dimensional matrix can be traced back to known, mostly named numbers (Re, Pr, Sc). The only new dimensionless number here is the Damköhler number Da which will be discussed later. The numbers obtained together with the four anticipated trivial numbers give the following dimensional-analytic framework:

$\{l/d,\ c_{out}/c_o,\ E/RT_o,\ \Delta T/T_o,\ k_o\tau,\ Re,\ Sc,\ Pr,\ Da\}$

The Π space under consideration is described completely by this Π set.

It is obvious that when scaling up chemical reactors it is generally unacceptable to change the reaction temperature T_o because this would impede the reaction course (and hence k_o) or at least the selectivity of the reaction. For the same reason, it is not possible to vary the physical or chemical properties of the reaction partners.

If scale-up of the tubular reactor of the given geometry (L/d = idem) is performed at T_o, $\Delta T/T_o$ = idem, taking account of these restrictions, the kinetic and material numbers E/RT_o, Da, Sc, Pr remain unchanged. Therefore, to attain a specified degree of conversion c_{out}/c_o = idem, it is only necessary to ensure that the other two numbers $Re = v\,d\,\rho/\eta$ and $k_o\tau \equiv k_o l/v$ are adjusted in such a way that they remain *idem*. However, it is immediately clear that this is an impossibility in the case of l/d = idem because $v\,d \rightarrow v\,l$ = idem *and* l/v = idem cannot be fulfilled simultaneously in the tubular reactor! (This problem does not exist in the mixing vessel, see Example C2).

With reference to scale-up of a *tubular reactor*, the problem is to increase the flowrate $q \sim v\,d^2$ by a factor n (not to be confused with the scale μ!) while retaining the chemical efficiency (yield, degree of conversion, selectivity, etc.):

$$q_H = n\,q_M \qquad \rightarrow \qquad v_H\,d_H^2 = n\,v_M\,d_M^2.$$

How does this demand react with the conditions Re = idem and $k_o\tau$ = idem?

$Re \sim v\,d$ = idem: $\quad d_H = n\,d_M; \quad v_H = v_M/n; \quad$ it follows that
$V_H = n^3\,V_M \quad$ and hence $\quad \tau_H = n^2\,\tau_M \quad (\tau = V/q)$.

The volume of the full-scale facility is n^3 times larger than that of the model. However, the flow through it is only $n\,q_M$, consequently the residence time τ is n^2 times larger. From $k_o\tau$ = idem it follows that the reaction rate constant k_o in the prototype would be smaller by a factor of n^2 and this contradicts the precondition stated above!

At this point, one will ask whether the demand for *Re* = *idem* is justified or not. In the last instance we are dealing with a fast reaction if we have selected a tubular reactor and the flow through it will certainly be turbulent. It is well known that Re only has a slight influence in the turbulent flow regime!

2. Heterogeneous catalytic reaction of the 1st order

Let us now consider a "catalytic packed bed reactor", i.e. a tubular reactor filled with a grained catalyst through which the gas mixture flows. With the diameter of the catalyst grain d_c, an additional dimensionless number d_c/d is added to the Π space; the Reynolds number is now expediently formed with d_c. The reaction rate is related to the unit of the bulk volume and characterized by an effective reaction rate constant $k_{o,\,eff} \equiv k^*$. λ_{eff} now indicates the thermal conductivity of the gas/bulk solids system too. Diffusion can be considered as negligible in this case (therefore Sc is not relevant). The complete Π space is therefore:

$$\{c_{out}/c_o,\ l/d,\ d_c/d\ ,\ E/RT_o,\ \Delta T/T_o,\ k^*\tau,\ Re,\ Pr,\ Da\}$$

Since the diameter of the catalyst grain has a considerable influence on the reaction rate, its variation will not be permitted during scale-up; this means that the geometry will *inevitably* be violated by $d_c/d \neq$ idem. Therefore, scale-up of the tubular reactor filled with catalyst is at best possible through adherence to partial similarity whereby it is necessary to check whether violation of geometric similarity alone is enough to guarantee scale-up.

The scale-up problem under discussion is completely outlined by the given Π space but it can be considered in greater depth by compiling fundamental differential equations which mathematically formulate the conditions for preservation of mass, impulse and energy (c.f. Fig.2 on p.24 , case D).

G. Damköhler [18] used this possibility to draw up *Navier-Stockes* differential equations of the mass and heat transfer for the case of an adiabatic reaction. Analytical solution of these differential equations is not possible. However, if they are made dimensionless, it becomes apparent that the Π space is formed by the five dimensionless numbers listed below:

$$\mathrm{Re} \equiv \frac{v\, l\, \rho}{\eta}$$

$$\mathrm{I} \equiv \frac{k\, l}{v} \quad \rightarrow \quad k\, \tau \text{ (for the pipe)}$$

$$\mathrm{II} \equiv \frac{k\, l^2}{\mathcal{D}} = \frac{k\, l}{v}\, \frac{v\, l}{\nu}\, \frac{\nu}{\mathcal{D}} = k\tau\, \mathrm{Re}\, \mathrm{Sc}$$

$$\mathrm{III} \equiv \frac{k\, c_o \Delta H_R\, l}{\rho\, c_p\, T_o\, v} = \frac{c_o \Delta H_R}{\rho\, c_p\, T_o}\, \frac{k\, l}{v} = \mathrm{Da}\, k\tau$$

$$\mathrm{IV} \equiv \frac{k\, c_o \Delta H_R\, l^2}{\lambda\, T_o} = \frac{c_o \Delta H_R}{\rho\, c_p\, T_o}\, \frac{k\, l}{v}\, \frac{v\, l\, \rho}{\eta}\, \frac{c_p\, \eta}{\lambda} = \mathrm{Da}\, k\tau\, \mathrm{Re}\, \mathrm{Pr}$$

Although *Damköhler* traced numbers I to IV back to the above combinations of named dimensionless numbers known at that time, numbers I to IV have come to be known as the four Damköhler numbers $\mathrm{Da_I}$ to $\mathrm{Da_{IV}}$ in chemical literature. We will *not* identify them in this way, instead we will refer only to the new additional reaction-kinetic Π variable

$$\boxed{\frac{c_o \Delta H_R}{\rho\, c_p\, T_o} \equiv \mathrm{Da}}$$

as the Damköhler number Da.

In fact, the advantage of the combinations of numbers obtained above by making differential equations dimensionless over those combinations delivered by the dimensional analysis is that they characterize certain types of mass and heat transfer respectively. For example III represents the ratio of the reaction heat to heat removal by convection while IV expresses the ratio of the reaction heat to heat removal by conduction.

G. Damköhler bases his analysis of the scale-up problem relating to the catalytic tubular reactor on the following Π set ($\mathcal{D}$ and hence number combination II are not relevant):

$$\left\{ \frac{l}{d}, \frac{d_c}{d}, \frac{v\, d_c\, \rho}{\eta}, \frac{k^*\, l}{v}, \frac{k\, c_o \Delta H_R\, l}{\rho\, c_p\, T_o\, v}, \frac{k\, c_o \Delta H_R\, l^2}{\lambda\, T_o} \right\}$$

Re I III IV

He knows that he may not vary the temperature T_o and d_c if he does not want to risk influencing the chemical course of the reaction. Consequently, as already mentioned, geometric similarity is inevitably violated during scale-up on account of $d_c/d \neq$ idem. *Damköhler* is therefore prepared to waive adherence to l/d = idem as well. However, he points out that this will necessarily lead to consequences for heat transfer behaviour. In this case, he uses the *hypothesis* that thermal similarity is guaranteed if the ratio of IV to III (heat conduction through the tube wall to heat removal by convection) is kept equal:

$$\frac{IV}{III} = \frac{k^*\, c_o \Delta H_R\, d^2}{\lambda\, T_o} \; \frac{\rho\, c_p\, T_o\, v}{k^*\, c_o \Delta H_R\, l} = \frac{\rho\, c_p\, v\, d^2}{\lambda\, l} \equiv Pe = idem.$$

Scale-up must therefore be effected in the following Π space:
$\{k^*\tau, Re, Pe\}$. It then follows that:

$Re = idem \rightarrow v = idem$
$k^*\tau = k^*\, l/v = idem \rightarrow l = idem$
$Pe = idem \rightarrow \boxed{d = idem}$.

The requirement d = idem makes a change in scale impossible.

Result: Abandoning geometric similarity does not suffice to guarantee chemical similarity (T_o and hence k^* = idem).

Damköhler now proposes abandoning not only geometric but also fluid dynamic similarity (Re = irrelevant) and depending exclusively on thermal and reaction similarity. This means that, apart from $k^*\tau$, only III and IV must be kept constant. The Π space is then:

$$\left\{ \frac{k^*\, l}{v}, \frac{k\, c_o \Delta H_R\, l}{\rho\, c_p\, T_o\, v}, \frac{k\, c_o \Delta H_R\, d^2}{\lambda\, T_o} \right\}.$$

Since, according to *Damköhler*, λ_{eff} is approximately proportional to the flowrate in the turbulent flow regime and, as from a certain small d_k/d

ratio, independent of this, the following scale-up rules result from the above Π space:

$Da_I \equiv k^*\tau \rightarrow \quad l/v = \text{idem},$

$Da_{III} \quad \rightarrow \quad l/v = \text{idem}$

$Da_{IV} \quad \rightarrow \quad d^2/\lambda_{eff} \sim d^2/v = \text{idem}$

In conjunction with the flow rate equation $q_H = n\, q_M$, it follows that

$Da_{IV} = \text{idem} \quad \rightarrow \quad d_H = n^{1/4}\, d_M \quad \text{and} \quad v_H = n^{1/2}\, v_M.$

$Da_I = \text{idem}$ as well as $Da_{III} = \text{idem} \quad \rightarrow \quad l_H = n^{1/2}\, l_M$

Consequently

$\boxed{(l/d)_H = n^{1/4}\,(l/d)_M}$ and in conjunction with $\Delta p \sim l\, v^2$ the following is valid

$\boxed{\Delta p_H = n^{3/2}\, \Delta p_M}$.

If one assumes that both hypotheses (irrelevance of both geometric and fluid dynamic similarity) are applicable, the result shows that the prototype would be linked with a pressure increase which would not only incur costs but could also have an unfavourable effect on the reaction. Bundles of small contact tubes are more economic.

Scale-up at partial similarity was discussed in Sect. 3.3 (p.42). It was pointed out that various strategies exist and Froude's method which is based on *dividing the process into parts* that can be investigated separately was presented.

Example C1 illustrates a further possibility for dealing with partial similarity. It is based on *deliberately abandoning certain similarity criteria* and theoretically and/or practically checking the effects on the entire process. *Damköhler*'s example which is presented here convincingly demonstrates the valuable information relating to scale-up of a complex chemical process which can be deduced on the basis of theoretical considerations alone if the principles relating to the theory of similarity are used consistently.

Damköhler himself, however, seems to have been very disappointed with the result of his study. His conclusions in [18] as quoted below cannot be interpreted any other way:

> "Although it is basically possible to apply the theory of similarity to chemical processes and to scale up one of these processes in such a way that geometric, fluid dynamic, thermal and reaction-kinetic similarity is retained to a greater or lesser extent, these transformation processes are only of limited importance. They may be quite useful for increasing equipment performance two to five-fold but hardly to much larger amounts. This circumstance is of importance since *it is more or less equivalent to practical failure of the theory of similarity*. This, however, was not to be expected from the beginning, especially in view of the fact that the theory of similarity proved itself brilliantly in the solution of other heat transfer problems where no additional chemical conditions had to be fulfilled."

The results of his studies and the efficiency of methods based on the theory of similarity have been assessed differently by posterity. If the method shows that scale-up is not possible, this by no means points to failure of the method but rather is a valuable indication of the given facts!

Example C2:

Influence of back-mixing (macromixing) on the degree of conversion in continuous chemical reaction operation

This example has almost nothing to do with dimensional analysis and scale-up in chemical reaction engineering. Nevertheless I felt that it was necessary to include it in this section for two reasons. *First* it shows the only correct method of combining the course of a chemical reaction with the fluid dynamics in continuous chemical reactors while focussing on their residence time distributions instead of on l/d and Re = idem (see the introduction to Example C1). From this point of view, these achievements of chemical reaction engineering of the 1950/60´s represent the straightforward continuation of the work begun by *G.Damköhler*.

Secondly, this example is the counterpart to Example C3: It introduces the scaling-up data for mixing vessels in series which consider the interaction of *macro-mixing* (back-mixing) and conversion rate while Example 3 deals with the interaction of *micro-mixing* and selectivity!

Example C1, Sect. 1 showed that, in the case of the *tubular reactor*, scale-up cannot be effected while retaining total similarity (l/d, Re = idem). In the *mixing vessel*, in contrast, the limitation l/d = idem is superfluous and the numerical value of Re is also insignificant in the turbulent flow regime.

When dealing with a continuous reaction process, greater importance is attached to the time the individual volume elements of the throughput have actually remained in the reactor. This depends on the nature of *macromixing* in the reaction chamber. *Back-mixing* of the reaction partners in the reactor not only has a considerable influence on the conversion rate, it can greatly affect the selectivity of the reaction too; see Example C3. A theoretical distinction is therefore made between an ideal tubular reactor (plug flow, no back-mixing) and an ideal mixing vessel (mixing time θ much shorter than the mean residence time τ : complete back-mixing). There are smooth transitions between these two extreme conditions: If one changes from a stirring vessel to N mixing vessels in series, one obtains a tubular reactor at $N = \infty$. On the other hand, a tubular reactor becomes a mixing vessel if the axial diffusion $\mathfrak{D}_{ax}$ is increased substantially.

This consideration clearly shows that, in a continuous reaction process, back-mixing and its suppression and hence $\mathfrak{D}_{ax}$ and N respectively are much more important than the flow condition which is described solely by the Re number. This new quality is identified by the so-called *residence time distribution characteristic* of the reactor which quantitatively indicates which fraction φ of the throughput actually remains for which period of time t in the reactor with the mean residence time τ : $\varphi = f(t/\tau, N)$, see Fig. B 2.1. (By the way, the correlation between $\mathfrak{D}_{ax}$ resp. $Bo_{ax} \equiv v\, l/\mathfrak{D}_{ax}$ and N is known for various reactor types[30]. $Bo_{ax} = 0$ stands for the ideally back-mixed vessel with N = 1 while $Bo_{ax} = 1$ stands for the plug-flow reactor with $N = \infty$.)

---

30) D. W. van Krevelen, Chem.- Ing. Tech. **30** (1958) 9, 553/559
J. Pawlowski, Chem.-Ing.- Tech. **34** (1962) , 628/631

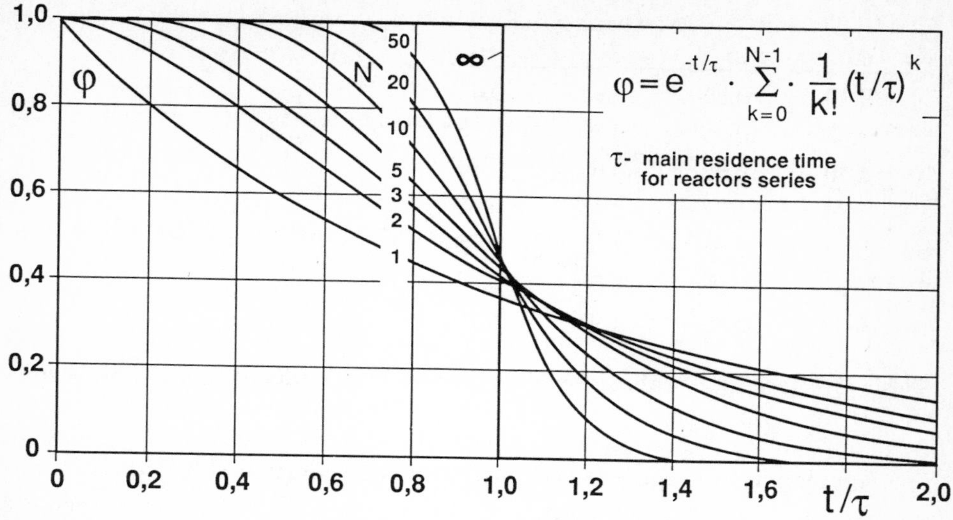

Fig. C 2.1: Theoretically calculated residence time distribution charac teristics for equal-sized back-mixed reactors in series

The relationship given in Fig. C 2.1 can be determined experimentally using so-called displacement marking: At the start of the experiment, the total reactor volume is marked (e.g., with common salt or dye). It is then displaced by a non-marked throughput and the concentration of the marking substance is measured in the outlet.

The relationship $\varphi = f(t/\tau)$ can e.g. be used to calculate how long the mean residence time of individual mixing vessels (with equal volumes) must be to ensure that P% of the throughput remains in them for an effective *minimum time* t. To give a fast answer to this interesting question, Table C 2.1 presents the numerical values of $t/\tau_{step} = t\,q/V_{step}$ for the step numbers N = 1 to 10 and the percentages of P = 90 to 99%.

Example: How large must the mixing vessels of a series of N = 3 be to ensure that 95% of the throughput of $q = 10\ m^3/h$ remains in the reactor cascade at least $t_{min} = 0.5$ h? The answer is:

$t/\tau_{step} = t\,q/V_{step} = 0.82$. It follows that:

$V_{step} = t\,q/0.82 = (0.5\ h \times 10\ m^3/h)/0.82 = 6.1\ m^3$. $V_{series} = 18.3\ m^3$.

N	P [%] 90	95	97	99
1	0.10	0.05	0.03	0.01
2	0.53	0.35	0.26	0.15
3	1.10	0.82	0.66	0.44
4	1.74	1.37	1.15	0.82
6	3.15	2.61	2.30	1.78
8	4.65	3.98	3.58	2.90
10	6.22	5.42	4.95	4.13

Table C 2.1: t/τ_{step} - values for N equal-sized back-mixed reactors in series

The connection between the residence time distribution characteristic of a reactor and the temporal course of the reaction provides the respective conversion rate. The mass balance of a reactor is described by the following differential equation:

$$dn/dt = -q\,dc + r\,dV$$

where: n – mol number of the passing reaction partner, q – liquid throughput, r – reaction rate, V – reactor volume.

Integration of this equation under the boundary conditions:

a) steady-state operating condition ($dn/dt = 0$) and

b) reaction of the 1st order ($r = -k\,c$)

gives the correlation for the ideal *plug-flow reactor*:

$$\frac{c_{out}}{c_o} = e^{-kt} \quad \text{resp.} \quad X = 1 - e^{-kt},$$

whereby the conversion rate X is defined as $X = 1 - c_{out}/c_o$.

In contrast, the correlation for the ideal *back-mixed reactor* (c and hence $r = -k\,c = const$) gives the following equation:

$$X = 1 - \frac{1}{1 + k\tau}.$$

This correlation is presented in the form of a working sheet in Fig. C 2.2.

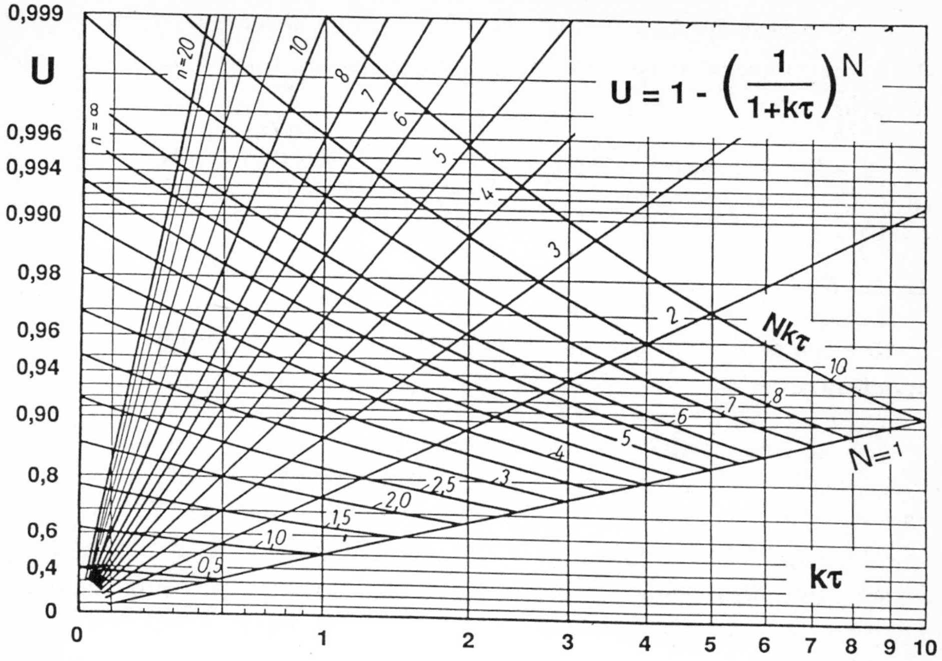

Fig. C 2.2: Working sheet for determination of volumes of N equal-sized back-mixed reactors in series for a 1st order chemical reaction (H. Hofmann, PhD-thesis, TH Darmstadt 1954)

Application example: Even without knowledge of the reaction rate constant k, the diagram shows how subdivision of the reaction volume into steps lowers the necessary total mean residence time $N\tau$. In order to attain a 99% conversion rate ($X = 0.99$) the following applies:

N	$k\tau$	$Nk\tau$
2	9.0	18
3	3.7	11
5	1.5	7.5
10	0.6	6

A 3-step cascade requires only 11/18 = 0.6 → 60% of the total volume of a 2-step cascade, while a 10-step cascade needs only 6/18 = 1/3 of the volume of the 2-step one.

If a 7-step cascade is operated at $k\tau = 1$ with an end conversion rate of $X = 0.99$, the conversion rate in the individual steps can be read from the working sheet:

$N = 1 \rightarrow X = 0.5$; $N = 3 \rightarrow X = 0.87$; $N = 5 \rightarrow X = 0.97$; etc.

Example C 3:

Influence of micro-mixing on selectivity in a continuous chemical reaction process

As a result of subdivision of the reactor volume into N equal-sized back-mixed reactors in series, the residence time distribution characteristic of the throughput is constricted and consequently the conversion rate of the continuous chemical reaction increased; Example C 2. For a homogeneous, simple, irreversible chemical reaction, the only requirement imposed on the mixing intensity in the individual steps is that the mixing time θ must be much smaller than the average residence time τ (so-called ideally back-mixed vessel).

In the case of a competitive consecutive reaction

$$A + B \xrightarrow{k_1} P \quad \text{(desired)}$$

$$P + B \xrightarrow{k_2} R \quad \text{(undesired)}$$

however, the demand for constriction of the residence time distribution - which implies a specific regulation for bulk back-mixing (macro-mixing) - is not sufficient to guarantee selectivity with reference to the desired reaction product P. Intensive mixing of the reaction components on a microscopic scale (so-called micro-mixing) is additionally required for this purpose in order to shorten the diffusion this purpose in order to shorten the diffusion paths and allow the reaction to take place immediately.

There are many reactions of this type. Best known are certain chlorinations and phosgenations as well as the azo reactions. The selectivity of one of these (conversion of α-naphthol with diazo sulphanilic acid into a simple coupled (desired) and twice coupled product) is used to quantitatively determine micro-mixing on a "molecular scale". In this context, *J.R. Bourne* (see e.g. [B26]) relates selectivity S to the non-desired by-product R and defines it with P and R because only these two dyes can be detected analytically:

$$S_R \equiv \frac{2R}{2R + P} = \frac{\text{mols B consumed for formation of R}}{\text{total mols B consumed}} \;.$$

The reaction-engineering task will be to *minimize* the quantity S_R.

The azo reaction under consideration is very fast with reference to P formation and very slow with reference to R formation ($k_1 \gg k_2$). Performance of this reaction in a tubular reactor which exhibits ideal plug-flow characteristics with regard to macro-mixing (no back-mixing) is recommended. Fast micro-mixing of the reaction partners is taken care of by a propulsion jet nozzle which impinges the smaller volume flow of B into the larger volume flow of A (stoichiometrically surplus component).

In considering this example using dimensional analysis, we will start with the following relevance list:

target quantity:	selectivity S_R
geom. parameter:	tube diameter d (as the characteristic length)
physical properties:	$c_A, c_B; \rho_A, \rho_B; \nu_A, \nu_B; \mathcal{D}$.
process parameters:	$k_1, k_2; q_A, q_B$.

(k_i $[L^3 T^{-1} N^{-1}]$ - reaction rate constants of reactions of the 2nd order; isothermal reaction process is presupposed)

After formation of the trivial numbers:
c_A/c_B ; ρ_A/ρ_B ; ν_A/ν_B ; $\nu/\mathcal{D}$; k_1/k_2 ; q_A/q_B
the quantities $\{d, c_B, \rho_B, \mathcal{D}, k_2, q_B\}$ remain. Of these, only the density ρ_B contains the primary quantity mass and must therefore be deleted (the density is included in ν !) Only c_B and k_2 contain the primary quantity mol; this is eliminated in the combination $c_B k_2$. The residual relevance list is then $\{d, \mathcal{D}, c_B k_2, q_B\}$. The dimensional matrix with the rank 2 supplies two dimensionless numbers:

$$Re \equiv \frac{q_B}{\nu\, d} \quad \text{and} \quad Da_{II} \equiv \frac{d^2\, c_B k_2}{\mathcal{D}}$$

(Da_{II} - Damköhler number II for the reaction of the 2nd order).

The complete Π set of 9 numbers

$$\boxed{\{S\,,\,c_A/c_B\,,\,\rho_A/\rho_B\,,\,v_A/v_B\,,\,v/\mathcal{D}\,,\,k_1/k_2\,,\,q_A/q_B\,,\,Re\,,\,Da_{II}\}}$$

is reduced for the given reaction in the given material system (ρ_A/ρ_B, v_A/v_B, $v/\mathcal{D}$, k_1/k_2 = const) to the dependence

$$\boxed{S = f\,(c_A/c_B\,,\,q_A/q_B\,,\,Re\,,\,Da_{II})}\,.$$

A fast chemical reaction is characterized by the fact that it is limited with regard to mass transfer. In order to shorten the diffusion paths and reduce the size of the segregated fluid bales in the direction of the so-called Kolmogoroff length λ

$$\lambda = \left(\frac{v^3}{\varepsilon}\right)^{1/4} \qquad \varepsilon = P/\rho V \text{ - mass-related stirrer power}$$

a high mixing power ε must be dissipated in the reaction volume. In the tube, this mixing power has to be related to the liquid throughput: P/q. Since the power of the propulsion jet nozzle is given as $P = q\,\Delta p$, it follows that $P/q = \Delta p$.

If the liquid throughput q is replaced by the throughput-related mixing power $P/q = \Delta p$ in the above relevance list, the characteristic linear dimension tube diameter d must be deleted as irrelevant on account of the intensity character of this variable.

In order to replace q by Δp, we will combine the Euler and Reynolds numbers accordingly

$$Eu\,Re^2 \equiv \frac{\Delta p\,d^4}{\rho\,q^2}\,\frac{q^2}{v^2\,d^2} \equiv \frac{\Delta p\,d^2}{\rho\,v^2}$$

and eliminate d from the resulting Π number using Da_{II}:

$$Eu\,Re^2\,Da_{II}^{-2}\,Sc \equiv \frac{\Delta p\,d^2}{\rho\,v^2}\,\frac{\mathcal{D}}{d^2\,c_B k_2}\,\frac{v}{\mathcal{D}} \equiv \frac{\Delta p}{\eta\,c_B k_2} \equiv \Psi$$

The resulting Π set which will describe the selectivity S_R of a fast chemical reaction in the tubular reactor with the nozzle as mixing device, is then as follows:

$$\boxed{S_R = f(\Psi, c_A/c_B, q_A/q_B)}.$$

Laboratory tests were performed in glass tubes (d = 3 - 25 mm) with converging nozzles manufactured by the Schlick company (d´ = 0.3 - 1.6 mm). q_B formed the propulsion jet. A temperature of 20 ^oC, pH = 10 and mole ratio of $c_A q_A/c_B q_B = 1.05$ were kept constant so that only the relationship

$$S = f(\alpha \equiv q_A/q_B, \Psi)$$

could be investigated. However, test conditions which permitted the formation of a back-mixing eddy in the pipe reactor and hence back-mixing of R and B were also adjusted. (The back-mixing eddy forms when the free jet sucks more liquid from its environment than is supplied to it.)

The results of these tests are presented in Fig. C3.1. The analytical expressions for the two groups of curves are:

without back-mixing: $S_R = 1.3\,(\Psi\, c_B/c_A)^{-0.5}\,\alpha^{1.25} + 0.001$

with back-mixing: $S_R = 5.0\,(\Psi\, c_B/c_A)^{-0.5}\,\alpha^{1.25} + 0.005$

If these results are considered for the selectivity S_P with reference to the desired product P (in the figure the selectivity S_R of the *non*-desired by-product is illustrated!), it becomes apparent that this increases with increasing Ψ but decreases superproportionally with increasing throughput ratio α. This is because the propulsion jet power related to the *total* liquid throughput $\Delta p q_B/(q_A+q_B)$ decreases with increasing α.

If the kinetic parameters $k_1/k_2 = 7300/1.63 = 4480$ and the mol ratio $c_A q_A/c_B q_B = 1.05$ are known, it is possible to calculate the selectivities S_R which would result in the ideal plug-flow tubular reactor and in the completely back-mixed vessel. The corresponding values are 0.001 and 0.008 respectively. This is in good agreement with the results obtained for high values of $\Psi\, c_B/c_A$.

Tebel u. May [B 28] obtained the same information as that presented in Fig. C3.1 in the course of a multidimensional, numerical simulation of

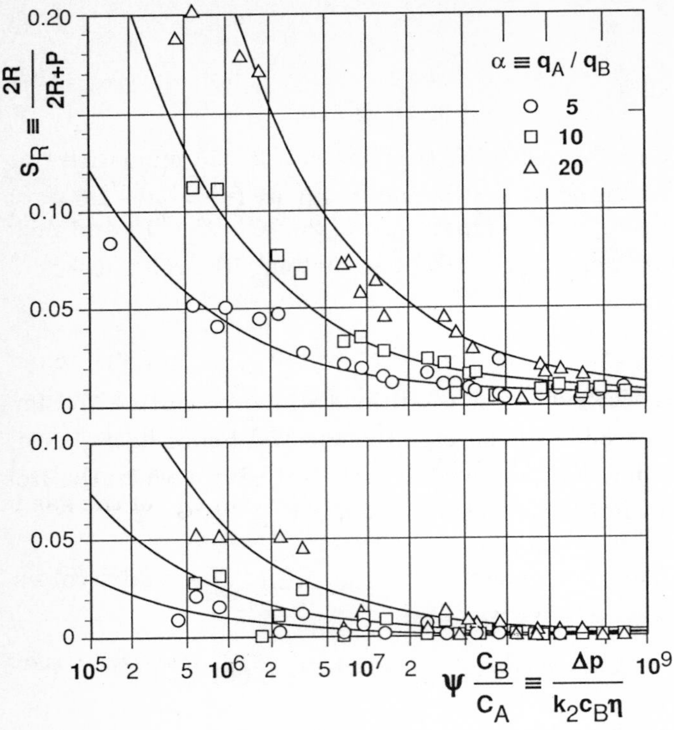

Fig. C.3.1: Selectivity S_R of the non-desired by-product R for a fast reaction taking place in the tubular reactor with Schlick nozzle as mixing device as a function of α and $\Psi(c_B/c_A)$. Taken from[B 27].
top:for back-mixing eddy formation *bottom*: for suppressed back-mixing

the turbulent flow field in which they only considered macro-mixing due to the convective terms of a turbulent flow in addition to the kinetic parameters of the reaction. In this case, it is indeed not necessary to explicitly consider the micromixing parameters (ε resp. Δp)!

[B 29] deals with the influence of different geometric conditions on $S_R = f(\varepsilon)$ for performance of the same reaction in the mixing vessel.

Example C4:

Mass transfer limitation of the reaction rate of fast chemical reactions in the heterogeneous material system gas/liquid

To allow a chemical reaction to take place between a gaseous and a liquid reaction partner, the gaseous component must first be dissolved (absorbed) in the liquid. In this case, the overall reaction rate will depend on the rates of the mass transfer and the chemical reaction step.

The so-called "two film theory" assumes the formation of laminar boundary layers on both sides of the interphase. Mass transfer through these layers can only be effected by means of diffusion, while the phase transition is immeasurably fast, Fig. C4.1. Consequently, equilibrium predominates in the interphase and the saturation concentration $c_G{}^*$ of the gas in the interphase (*) follows *Henry*'s law:

$$c_G{}^* = Hy\, p_G$$

(Hy - Henry coefficient, p_G - partial pressure of the gaseous reaction partner; indices: G - gas, L - liquid)

The two mass transfer coefficients k_G and k_L give the ratio of the respective diffusion coefficient $\mathfrak{D}_i$ to the respective boundary layer thickness x_i: $k_i \sim \mathfrak{D}_i/x_i$. However, since $k_G \gg k_L$, only the influence of k_L is taken into account in the following.

Technically important reactions of gases (A) with liquids (B) are mostly reactions of the 2nd order which take place according to the scheme [B 30]:

rate-governing:	$A + B \rightarrow C + E$
very fast consecutive reaction:	$E + (z - 1)B \rightarrow F$
stoichiometric overall reaction:	$A + zB \rightarrow C + F$

If the mass transfer of a gaseous reaction partner into the liquid is accompanied by a chemical reaction, the following case can occur depending on the reaction rate and the mobility of the reaction partners: The concentration of A is not only reduced to zero in the solution; in addition, the reaction front shifts from the bulk of the liquid to the liquid-side

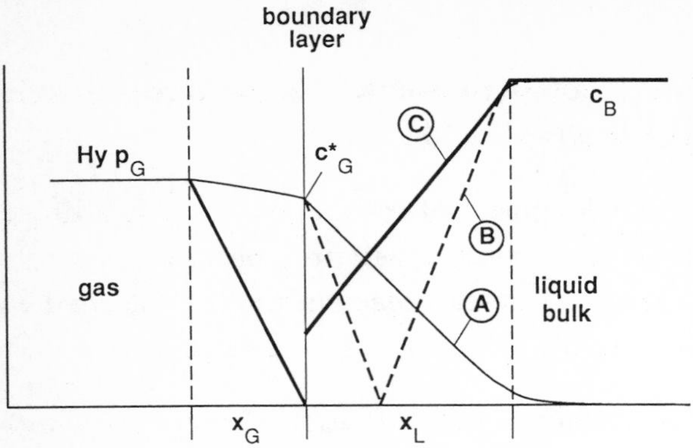

Fig. C4.1: Graphic depiction of the concentration profiles near the interphase for A) physical absorption, B) absorption with a fast chemical reaction, C) "chemisorption".

boundary layer, as a result of which this is apparently decreased to eliminated in a chemical way ("chemisorption"), see Fig. C4.1. This process increases the mass transfer coefficient by the "*enhancement* or *acceleration factor* m" as compared to its numerical value for purely physical absorption.

The target quantity m will depend on the parameters of mass transfer (k_L, $\mathfrak{D}_A$, $\mathfrak{D}_B$) and reaction kinetics (k_2, c_A^*, c_B). Apart from the trivial, obvious dimensionless numbers m, $\mathfrak{D}_A/\mathfrak{D}_B$, c_A^*/c_B four parameters remain which form a single additional dimensionless number, namely the *Hatta* number Hat (for the reaction of the 2nd order):

$$\mathrm{Hat} \equiv \sqrt{\mathfrak{D}_A k_2 c_B}/k_L.$$

If the numbers $\mathfrak{D}_A/\mathfrak{D}_B$ and c_A^*/c_B are formulated as a ratio of the diffusion currents $Z \equiv \mathfrak{D}_B c_B / z \mathfrak{D}_A c_A^*$, the resulting functional relationship is:

$$\boxed{m = f(\mathrm{Hat}, Z)} \; .$$

This Π relationship can be treated theoretically by assuming that the gradients of the diffusion rates of the two mass flows and the chemical reation rate are equal (quasi steady-state):

$$\mathfrak{D}_A \frac{d^2c_A^*}{dx^2} = \mathfrak{D}_B \frac{d^2c_B}{dx^2} = k\ c_A^*\ c_B\ .$$

This differential equation was numerically solved and graphically presented as a work sheet, Fig. C4.2, by *van Krevelen und Hoftijzer* [B 30] using the simplifying assumption of an idealized concentration profile.

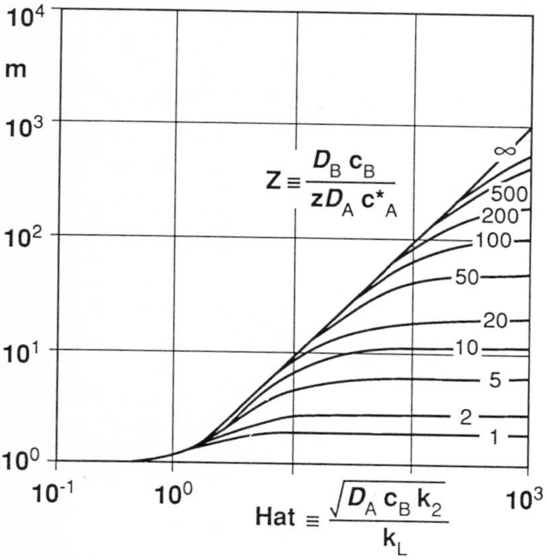

Fig. C4.2: Acceleration factor m in the case of gas absorption with subsequent chemical conversion as a function of the Hatta number Hat and the ratio of the diffusion currents Z.

The following three ranges can be differentiated:

a) Hat < 0,5: m = 1. In this case, physical absorption governs the rate. This can be increased as described above, see Examples B4 and B5.

b) Hat > 2; $c_B \gg c_A^* \rightarrow Z \approx \infty$: m increases directly proportionally to Hat. Since m was defined as m $g/(k_L\ c_A^*)$ (g - mass flow density G/A for physical absorption) it follows that

$$\frac{m\ g}{k_L\ c_A^*} = \frac{\sqrt{\mathfrak{D}_A\ k_2\ c_B}}{k_L}\ ; \quad m\ g = \sqrt{\mathfrak{D}_A\ k_2}\ c_A^*\ \sqrt{c_B}$$

From this relationship, it is apparent that in the case of chemisorption k_L is replaced by the kinetic expression $\sqrt{\mathfrak{D}_A k_2}$ which, in the given reaction system, depends only on the temperature.

c) With increasing k_2 and, subsequently, increasing Hatta number, the molecular mobility of B and hence Z become increasingly important. In this range, m is increasingly dependent on the mixing intensity in the reaction system which reduces the diffusion resistances.

(Similar relationships are found for reactions in the gas/solids system where pore diffusion is superimposed on the chemical reaction rate; see textbooks on chemical reaction engineering)

Important, named dimensionless numbers

Name	Symbol	Group	Remarks
A Mechanical Unit Operations			
Reynolds	Re	$v\,l/\nu$	$\nu \equiv \eta/\rho$
Froude	Fr	$v^2/(l\,g)$	
	Fr*	$v^2\,\rho/(l\,g\,\Delta\rho)$	$\equiv$ Fr $(\rho/\Delta\rho)$
Galilei	Ga	$g\,l^3/\nu^2$	$\equiv$ Re^2/Fr
Archimedes	Ar	$g\Delta\rho\,l^3/(\rho\,\nu^2)$	$\equiv$ Ga $(\rho/\Delta\rho)$
Euler	Eu	$p/(\rho\,v^2)$	
Newton	Ne	$F/(\rho\,v^2\,l^2)$	
		$P/(\rho\,v^3\,l^2)$	
Weber	We	$\rho\,v^2\,l/\sigma$	
Mach	Ma	v/v_s	
Knudsen	Kn	λ_m/l	
B Thermal Unit Operations (Heat transfer)			
Nusselt	Nu	$\alpha\,l/\lambda$	
Prandtl	Pr	ν/a	$a \equiv \lambda/(\rho\,c_p)$
Grashof	Gr	$\beta\Delta T\,g\,l^3/\nu^2$	$\equiv \beta\Delta T$ Ga
Fourier	Fo	$a\,t/l^2$	
Péclet	Pe	$v\,L/a$	$\equiv$ Re Pr
Rayleigh	Ra	$\beta\Delta T\,g\,l^3/(a\,\nu)$	$\equiv$ Gr Pr
Stanton	St	$\alpha/(v\,\rho\,c_p)$	$\equiv$ Nu/(Re Pr)
C Thermal Unit Operations (Mass transfer)			
Sherwood	Sh	$k\,l/D$	
Schmidt	Sc	ν/D	
Bodenstein	Bo	$v\,l/D_{ax}$	$\equiv$ Re Sc_{ax}
Lewis	Le	a/D	$\equiv$ Sc/Pr
Stanton	St	k/v	$\equiv$ Sh/(Re Sc)

D Chemical Reaction Engineering

Arrhenius	Arr	$E/(\mathcal{R}T)$	
Hatta	Hat	$(k_1 \mathfrak{D})^{1/2}/k_L$ $(k_2 c_2 \mathfrak{D})^{1/2}/k_L$	
Damköhler	Da	$\frac{c\, \Delta H_R}{c_p\, \rho\, T_o}$	proposal of this book
	Da_I	$k_1\, \tau$	
	Da_{II}	$k_1\, L^2/\mathfrak{D}$	$\equiv Da_I\, Bo = Da_I\, Re\, Sc$
	Da_{III}	$k_1\, \tau\, (\frac{c\, \Delta H_R}{c_p\, \rho\, T_o})$	$\equiv Da_I\, (\frac{c\, \Delta H_R}{c_p\, \rho\, T_o})$
	Da_{IV}	$\frac{k_1\, c\, \Delta H_R\, l^2}{\lambda\, T_o}$	$\equiv Da_I\, Re\, Pr\, (\frac{c\, \Delta H_R}{c_p\, \rho\, T_o})$

Legend

a	thermal diffusivity ($\equiv \lambda/\rho\, c$)
c, c_1, c_2	concentration
c_p	heat capacity at constant p
$\mathfrak{D}$	diffusivity
$\mathfrak{D}_{ax}$	eff. axial dispersion coefficient
E	activation energy
F	force
g	gravitational constant
ΔH_R	heat of reaction
k, k_L	mass transfer coeff. (index $_L$ - liquid side)
k_1, k_2	reaction rate constant (index: reaction order)
l	length
$p, \Delta p$	pressure, pressure difference
P	power
$\mathcal{R}$	universal gas constant
t	time
$T, \Delta T$	temperature, temperature difference
v	velocity

v_s	velocity of sound
α	heat transfer coefficient
β	temp. coeff. of density
ν	kinematic viscosity
η	dynamic viscosity
λ	heat conductivity
λ_m	molecular free path length
ρ, $\Delta\rho$	density, density difference
σ	surface tension
τ	residence time

Literature

A Treatment of single topics

[1] da Vinci, L., *Notebooks* (around **1500**); from [A3]

[2] Galilei, G., *Discorsi* (**1638**); German in Ostwalds Klassiker, Heft 11, 106/109

[3] Newton, I., *Principia* (**1687**); liber II, sectio VII, propositio 32

[4] Bertrand, R., Compt. Rend. **25** (1847),163; J.d'école polyt. (1848) 32,189

[5] Merrifield, C. W., Trans.Inst.Naval Arch. (London) **11** (1870), 80/93: *The experiments recently proposed on the resistance of ships*

[6] Reynolds, O., Philos. Trans. R. Soc. London **174** (1883), 935/982: *An experimental investigation of the circumstances whichdetermine whether the motion of water shall be direct or sinuous, and of the law of resistance in parallel channels*

[7] Baekeland, L. H., J. Ind. Eng. Chem. **8** (1916), 184/190: *Practical life as a complement to university education - medal address* (At the acceptance of the Perkin medal award)

[8] Görtler, H., ZAMM = Zschr. Angew. Math. Mech. **55** (1975), 3/8: *Zur Geschichte des Π-Theorems*

[9] Fourier, J. B. J., *Théorie analytique de la chaleur*, Paris **1822**

[10] Helmholtz, H.v., Monatsber. Kgl. Preuß. Akad. Wiss. Berlin (**1873**), 501/514: *Über ein Theorem, geometrisch ähnliche Bewegungen flüssiger Körper betreffend, nebst Anwendung auf das Problem, Luftballons zu lenken*

[11] Rayleigh, Lord, Nature **95** (1915) No 2368 (March 18), 66/68: *The principle of similitude*

[12] Vaschy, A., *Traité d'électricité et de magnétisme*, tome I., Baudry et Cie, Paris **1890**

[13] Federmann, A., Annalen des Polytechnischen Instituts Peter der Große, St.Petersburg **16** (1911), 97/154: *Über einige allgemeine Integrationsmethoden der partiellen Differentialgleichungen erster Ordnung*

[14] Riabouchinsky, D., L' aérophile **19** (1911), 407/408:
Méthode des variables de dimension zéro

[15] Buckingham, E., Physical Review, New York; 2nd Series **4** (1914) 4, 345/376:
On physically similar systems;Illustrations of the use of dimensional equations

[16] Rayleigh, Lord, Proc. Roy. Soc. London **66** (1899-1900), 68/74:
On the Viscosity of Argon as affected by Temperature

[17] Froude, W., Trans. Inst. Naval Arch. (London) **15** (1874), 36/73:
On experiments with H.M.S. "Greyhound"

[18] Damköhler, G., Z. f. Elektrochemie **42** (1936), 846/862:
Einflüsse der Strömung, Diffusion und des Wärmeübergarges auf die Leistung von Reaktionsöfen

[19] Zlokarnik, M., Chem.-Ing.-Tech. **55** (1982) 5, 363/372:
Modellübertragung in der Verfahrenstechnik
Ger. Chem. Eng.**7** (1984), 150/159:
Scale-up in Process Engineering

[20] Zlokarnik, M., Chem.-Ing.-Tech. **57** (1985) 5, 410/416:
Modellübertragung bei partieller Ähnlichkeit
Int. Chem. Eng.**27** (1987) 1, 1/9:
Scale-up under conditions of partial similarity

[21] Pawlowski, J., Verfahrenstechnik **8** (1974) 9, 269/272:
Modellversuche an Newtonschen Flüssigkeiten mit temperaturabhängiger Viskosität

[22] Pawlowski, J., Rheologica Acta **6** (1967) 1, 54/61:
Zur Theorie der Ähnlichkeitsübertragung bei Transportvorgängen in nicht-Newtonschen Stoffen
AIChE Journal **15** (1969) 2, 303/305:
Relationships Between Process Equations for Processes in Connection With Newtonian and Non-Newtonian Substances

[23] Henzler, H.-J., Chem.-Ing.-Tech. **60** (1988) 1, 1/8:
Rheologische Stoffeigenschaften - Erklärung, Messung, Erfassung und Bedeutung

[24] Henzler, H.-J. and Schäfer, E.E., Chem.-Ing.-Tech. **59** (1987) 940/944:
Viskose und elastische Eigenschaften von Fermentationslösungen

B Books and general treatises

[A 1] Weber, M., Jahrb. Schiffbautech. Ges. **20** (1919), 355/477: *Die Grundlagen der Ähnlichkeitsmechanik und ihre Verwertung bei Modellversuchen*

[A 2] Weber, M., Jahrb. Schiffbautech. Ges. **31** (1930), 274/388 : *Das allgemeine Ähnlichkeitsprinzip der Physik und sein Zusammenhang mit der Dimensionslehre und der Modellwissenschaft*

[A 3] Johnstone, R. E. und Thring M. W.: *Pilot Plants,Models, and Scale-up Methods in Chemical Engineering* McGraw-Hill Co., New York **1957**

[A 4] Gröber-Erk-Grigull: *Die Grundgesetze der Wärmeübertragung* Springer-Verlag, Berlin/Göttingen/Heidelberg **1963**; S.159ff

[A 5] Bridgman, P. W.: *Dimensional analysis*
Yale University Press, New Haven, 1922, 1931, 1951;
Reprinted by AMS Press, New York **1978**
German translation by H. Holl:
"Theorie der physikalischen Dimensionen - Ähnlichkeitsbetrachtungen in der Physik" ;
Verlag B. G. Teubner, Leipzig und Berlin **1932**

[A 6] Langhaar, H. L., *Dimensional Analysis and Theory of Models*
John Wiley & Sons,Inc., New York **1951**
Reprinted by R. E. Krieger Publ.Co.Inc.,Huntington,N.Y. **1980**

[A 7] Pankhurst, R. C.: *Dimensional analysis and scale factors*
Chapman and Hall Ltd, London **1964**

[A 8] Sedov, L. I.: *Similarity and Dimensional Methods in Mechanics*
Original Russian publication: Moskow 1943
English translation at Academic Press, New York **1959**

[A 9] Pawlowski, J.: *Die Ähnlichkeitstheorie in der physikalisch-technischen Forschung - Grundlagen und Anwendungen*
Springer-Verlag Berlin-Heidelberg-New York **1971**

[A 10] Görtler, H.: *Dimensionsanalyse - Theorie der physikalischen Dimensionen mit Anwendungen*
Springer Verlag Berlin-Heidelberg-New York **1975**

[A 11] Haeder, W. and Gärtner, E.: *Die gesetzlichen Einheiten in der Technik*; issued by the Deutsche Normenausschuß (DNA), Berlin; Beuth-Vertrieb GmbH, Berlin 30 - Köln - Frankfurt/Main

[A 12] Matz, W.: *Anwendung des Ähnlichkeitsgrundsatzes in der Verfahrenstechnik*
Springer-Verlag Berlin/Göttingen/Heidelberg **1954**

[A 13] Pawlowski, J.: *Veränderliche Stoffgrößen in der Ähnlichkeitstheorie.* H.R.Sauerländer & Co., Aarau **1991**

[A 14] Pawlowski, J.: *Einwellen-Schnecken;Förder-,Homogenisier- und Wärmeaustausch-Verhalten*
H.R.Sauerländer & Co., Aarau **1990**

[A 15] Zierep, J.: *Ähnlichkeitsgesetze und Modellregeln der Strömungslehre;* (Wissenschaft + Technik: pocket edition), G.Braun, Karlsruhe **1982**

C Examples of Application

[B 1] Zlokarnik, M., Chem.-Ing.-Tech. **42** (1970) 15, 1009/1011:
Einfluß der Dichte- und Zähigkeitsunterschiede auf die Mischzeit beim Homogenisieren von Flüssigkeitsgemischen

[B 2] Zlokarnik, M., Korrespondenz Abwasser **32** (1985) 7, 598/603:
Neue Flotationstechniken zur Abtrennung und Eindickung von Klärschlamm bei der biologischen Abwasserreinigung
Teil 2: Entgasungsflotation

[B 3] Stanton, T. E. and Pannell, J. R., Phil.Trans.Roy.Soc.London **214** (1914), 199/225:
Similarity of Motion in Relation to the Surface Friction of Fluids

[B 4] Nikuradze, J., VDI-Forschungsheft **361**, Juli/August 1933:
Strömungsgesetze in rauhen Rohren

[B 5] Zlokarnik, M., Chem.-Ing.-Tech. **39** (1967) 9/10, 539/548:
Eignung von Rührern zum Homogenisieren von Flüssigkeitsgemischen

[B 6] Zlokarnik, M., Chem.-Ing.-Tech. **45** (1973) 10a, 689/692:
Rührleistung in begaster Flüssigkeit

[B 7] Zlokarnik, M., and Judat, H.: *"Stirring"* in Ullmann´s Encyclopedia of Industrial Chemistry, Vol.B 2, Chapt. 25, p. 1/33, VCH Verlagsgesellschaft, D-6940 Weinheim, **1988**

[B 8] Zlokarnik, M., Chem.-Ing.-Tech. **50** (1978) 9, 715:
Sorptions-Charakteristiken des Schlitzstrahlers in Abhängigkeit von den Koaleszenzbedingungen des Systems
Chem. Eng. Sci. **34** (1979) 10, 1265/1271
Sorption characteristics of slot injectors and their dependency on the coalescence behaviour of the system

[B 9] Zlokarnik, M., Chem.-Ing.-Tech. **40** (1968) 15, 765/768:
Homogenisieren von Flüssigkeiten durch aufsteigende Gasblasen

[B 10] Zlokarnik, M., Chem.-Ing.-Tech. **38** (1966) 3, 357/366:
Auslegung von Hohlrührern zur Flüssigkeitsbegasung. (Bestimmung des Gasdurchsatzes und der Wellenleistung)

[B 11] Müller, W. and Rumpf, H., Chem.-Ing.-Tech. **39** (1967) 5/6, 365/373:
Das Mischen von Pulvern in Mischern mit axialer Mischbewegung

[B 12] Zlokarnik, M., Chem.-Ing.-Tech. **43** (1971) 6, 329/335:
Eignung von Einlochböden als Gasverteiler in Blasensäulen

[B 13] Zlokarnik, M., Chem.-Ing.-Tech. **56** (1984) 11, 839/844:
Auslegung und Dimensionierung eines mechanischen Schaumzerstörers
Ger.Chem.Eng. **9** (1986) 5, 314/320:
Design and Scale-up of Mechanical Foam Breakers

[B 14] Zlokarnik, M., Chem.-Ing.-Tech. **41** (1969) 22, 1195/1202:
Wärmeübergang an der Wand eines Rührbehälters beim Kühlen und Heizen im Bereich $10^0 < Re < 10^5$

[B 15] Pawlowski, J. and Zlokarnik, M., Chem.-Ing.-Tech. **44** (1972) 16, 982/986:
Optimieren von Rührern für eine maximale Ableitung von Reaktionswärme

[B 16] Kast, W., Chem.-Ing.-Tech. **35** (1963) 11, 785/788:
Untersuchungen zum Wärmeübergang in Blasensäulen

[B 17] Zlokarnik, M., Chem.-Ing.-Tech. **38** (1966) 7, 717/723:
Auslegung von Hohlrührern zur Flüssigkeitsbegasung. (Ermittlung des erreichbaren Stoff- und Wärmeaustausches)

[B 18] Zlokarnik, M., Adv. Biochem. Eng. **8** (1978), 133/151: *Sorption Characteristics for Gas-Liquid Contacting in Mixing Vessels*

[B 19] Judat, H., Chem.-Ing.-Tech. **54** (1982) 7, 520/521: *Stoffaustausch Gas/Flüssigkeit im Rührkessel - eine kritische Bestandsaufnahme.* Ger. Chem. Eng. **5** (1982) 6, 357/363: *Gas/Liquid Mass Transfer in Stirred Vessels - A Critical Review*

[B 20] Zlokarnik, M., Adv. Biochem. Eng.**11** (1979), 157/180: *Scale-up of Surface Aerators for Waste Water Treatment*

[B 21] Zlokarnik, M., Korrespondenz Abwasser **27** (1980) 7, 14/21: *Eignung und Leistungsfähigkeit von Oberflächenbelüftern für biologische Abwasserreinigungsanlagen*

[B 22] Zlokarnik, M., Verfahrenstechnik 13 (**1978**) 7/8,601/604: *Die Leistungsfähigkeit optimierter Injektoren für den Sauerstoff-eintrag in biologischen Abwasserreinigungsanlagen*

[B 23] Zlokarnik, M., Biotechnology, (Editors: H.-J. Rehm und G. Reed), VCH Verlagsgesellschaft, Weinheim, Vol 2 (**1985**), p.537/569: *Tower-Shaped Reactors for Aerobic Biological Waste Water Treatment*

[B 26] Baldyga, J. and Bourne, J. R., *Principles of Micromixing,* in Encyclopedia of Fluid Mechanics (Ed.: Cheremisinoff) Vol.1, Chapt. 6, p. 148/195; Gulf Publ.Comp., Houston **1985**

[B 27] Korischem, B., Diplomarbeit Universität Dortmund **1987**, *Homogenisieren von Flüssigkeiten mit Düsen* (the results are published in [B 28])

[B 28] Tebel, K. H. a. May, H.-O., Chem.-Ing.-Tech.**60** (1988) 11, 912/913: *Der Freistrahlrohrreaktor - Ein effektives Reaktor-Design zur Unterdrückung von Selektivitätsverlusten durch schnelle, unerwünschte Folgereaktionen*

[B 29] Rice, R. W. a. Baud, R. E., AIChE J. **36** (1990) 2, 293/298: *The role of micromixing in ther scale-up of geometrically similar batch reactors*

[B 30] Krevelen, D. W. van and Hoftyzer, P. J., Chem.Eng.Sci **2** (1953) 4, 145/156: *Graphical design of gas-liquid reactors*

Index